Franklin Root Rathbun

Bright feathers

Some North American birds of beauty

Franklin Root Rathbun

Bright feathers
Some North American birds of beauty

ISBN/EAN: 9783337175122

Printed in Europe, USA, Canada, Australia, Japan

Cover: Foto ©berggeist007 / pixelio.de

More available books at **www.hansebooks.com**

BRIGHT FEATHERS

OR

Some North American

BIRDS of BEAUTY.

By Frank R. Rathbun.

Illustrated with Drawings made from Nature, and carefully Colored by Hand.

AUBURN, N. Y.
Published by the Author,
1880.

AUBURN, N. Y.
PRESS OF KNAPP & PECK,
118 Genesee Street.

To the

Memory of My Mother:

MARY LEONARD HAWLEY.

Whose admiration for Nature and her beauties,

early instilled into my heart an ardent love

for feathered forms, this work is

dedicated in grateful

remembrance

by

The Author.

Hic Opus

INTRODUCTION.

I N presenting this work to the public, the author deems it but reasonable that his patrons should be placed in full possession of the facts
and circumstances which have induced him to undertake so presuming
and (to him) so hazardous an enterprise.

When the publication of the "Revised List of the Birds of Central
New York" was an accomplished fact, many, into whose hands it was
placed, suggested to him the propriety of describing and illustrating
some, if not all of the birds contained therein. He felt that to truth-
fully and thoroughly execute the first proposition, was beyond his power
except in a few minor instances, and much less assuring of a promising
result than the latter suggestion. When, however, the horizon of a
doubtful success had been considered, and the material in his possession
consisting of the writings and assured observations of the ornithological
students of the last decade had been carefully scanned, he felt a growing
confidence that whenever his pen was blocked by a vacuitous doubt,
he could find relief therefor in referring to such publications as the
Nuttall Ornithological Bulletin, the Bulletins of the U. S. Geological

and Geographical Survey of the Territories, and the writings of such
authentic observers as Coues, Allen, Merriam, Brewster, Deane, and a
host of other like authorities, as well as to the Natural History Columns
of *Forest and Stream and Rod and Gun*, where many a note of value is
preserved.

Besides this, numerous requests have been made for drawings of our
native birds, (particularly of those whose beauty of form or plumage
rendered them more than ordinarily conspicuous), at nominal prices,
which the time necessarily consumed in their production would by no
means warrant. In this the initial number of the series, he has endeavored to present an example of how he intends to meet the wants enu-
merated, and however faulty the construction of the text may appear,
or however it may suggest to the reader a want of experience upon
the author's part, he can only assure those who favor his enterprise, that
a hearty and *practical* wish and encouragement for success therein will
tend to stimulate both his brush and pen and render the future of
"BRIGHT FEATHERS" more faithful in its delineations of bird-forms,
and more acceptable in its textual arrangement.

The series is intended to include ten or more species (in as many
parts) of such birds as are found in the North-eastern portion of the
United States, most attractive for their plumage colorations ; and the
plates will contain figures of the female as well as the male of the
species under advisement.

The figures are drawn from nature and are colored by hand, being earnest endeavors at truth, though the artist must confess that in comparing the subtle tinctures and deft painting of nature with his own attempts, he is full of discouragements and misgivings as to their success. He can only say, finally, that he has in the present part, as he will endeavor to do in the future of the work, (if carried to completion), tried earnestly to meet the wants already set forth, and exerted his best endeavor in the double capacity of artist and author to render his experiment at least a partial success.

Frank B. Rathbun

AUBURN, N. Y., July 1, 1880.

CARPODACUS PURPUREUS, (Gm.) Gray.

THE PURPLE FINCH. ♂·MALE . ♀·FEMALE . LIFE SIZE.

DRAWN FROM NATURE. BY FRANK R.RATHBUN.

BRIGHT FEATHERS

THE PURPLE FINCH.

CARPODACUS PURPUREUS, (GMELIN) GRAY.

PLATE I.—MALE AND FEMALE.

HIS most beautiful vocalist, has of late years excited the attention and curiosity of many people ; and the writer has had propounded to him many and varied descriptions of its plumage, its actions and its song, as a preface to an inquiry for its name, history and habits. For this reason alone, he has thought proper to prelude " BRIGHT FEATHERS," with as full and complete a description of its characteristics as he can command from personal observation and the manifold notes of other writers.

The manners and habits of this happy herald of balmy skies, are alone enough to command the attention of the most careless observer, even though his crimson garb be not brought into requisition to enhance his attractiveness. When, however, under the favorable auspices of the shimmering sunbeams of a welcome springtime his colorations are displayed to the view, and when these are supplemented by his sweet and wavering song, then the admirer of our native birds and the casual student of nature, halts and wonders if this handsome vocalist of the pulsing season be not a rare visitor, or some stray wanderer from the tropics.

Rare or infrequent he seldom becomes, for as far as my observations extend, (covering a period of some fifteen years,) he has been, in Central New York, a constant and comparatively abundant visitor.

Neither does he hail from a flowery land of sunny seasons, but upon the contrary comes to us from the bleak regions of the north ; bringing with him a glowing passion of which his garb is a fitting emblem ; helping and cheering his humbly clad mate in her toils, and brightening nature, already glad, with his attractive coat and ready tongue.

He comes to us sometimes a little earlier, but seldom later than the first weeks in March, and almost invariably the first intimation we have of his presence is the ripple of his sweet song from the sky above, where he flutters on his tremulous wings, or from the crowning twig of some lofty elm or maple tree, from whence his voice falls with peculiar and thrilling effect. His notes are very sweet and of continued strength, varying in their modulations and inflections, and I might almost say drawn out in harmonious airs without reiteration or repetition. I have found it almost impossible to attempt its score, and will not venture to write it.

The Purple Finch (as already intimated) " Sings on the wing," and from this fact alone has drawn upon itself the notice of many who would otherwise give it no attention. Seeking as it does, rather lofty places for its nestings, its love-makings and its musical efforts, its handsomely colored form is brought into view only when aided and supplemented by its closer proximity to the observer. It frequently happens that in its searchings for food among the briers and shrubs of our lawns and gardens, it will alight upon some lowly and convenient branch to plume itself and give utterance to a low, single and rather mellifluous *clink*, as if completely satisfied and happy with itself and the bursting life and odors about it. It is at such times and under such circumstances, the whole enhanced by a bright sun, that the Purple Finch displays his glory; and the fortunate observer cannot fail of being impressed with his beauty.

I will remember the first time that I ever saw this bird in close proximity; and his sudden advent upon my admiring gaze has served to stamp indelibly the place and circumstances upon my memory; so much so in fact, that whenever he is brought to my view, the whole phantom train of the intervening years is hurriedly encompassed, and I am again placed in the same spot, and engirdled by the same scenes and surroundings as those in which I first beheld his flaming beauty.

For some boyish misdemeanor, I think the emptying of my Mother's sugar candy or purloining from her honey store, I had been doomed to an imprisonment in a dark spare room, upon the first floor front of a pleasant country home in Northern Vermont. A few days previous to my unfortunate confinement, a poor stray waif of humanity, unknown, and with no evidence about him whereby he might have been traced, had in a solitary and adjacent wood been found hanging by the neck from a stout sapling; dead! He had used, to accomplish his purpose, a scarlet scarf or kerchief of some kind, and as we, a group of horrified urchins gazed upon his suspended form, the brighter color of the stifling band which seemed to enliven and warm up the supernatural scene, left an indelible impress upon my mind. From the time of the discovery of the inanimate form of the poor homeless Frenchman, (for such he was adjudged to be), to that of my pilferings, he had formed the principal subject of the conversation of my mates and myself; and around him had clustered all the horrors of youthful conceits and weird tales of ghostly lore. "How he looked!" "How his neck was lengthened by that scarlet woof!" "The place will, as a matter of course, and of a surety be haunted by his uneasy spirit, and we must henceforth avoid that portion of the wood!!" Such expressions, and such thoughts, were the burden of our associative hours; and my punishment, considering the complete isolation of my position, was greatly, if not cruelly enhanced thereby.

I had clandestinely and contrary to instructions, partially opened one of the green blinds of my prison enough to obtain a view of the small but pleasant lawn fronting my father's dwelling. Nature was in one of her most inviting moods; the voices of my boyish companions at a merry game of base ball, were borne to me upon one of her softest breathings, while the shadow of the paternal roof was gradually extending itself across the lawn and into the street. My youthful heart had yielded up its bitterness and resentment, to the terror inspired by ghostly meditations upon the self-sacrifice which had lately been enacted in the neighborhood, and was fast yielding vassalage to its complete sway, when a slight shadow flitting by the casement caused me to look up. For an instant my throat was choked, as I beheld what I imagined to be a *spot of swaying blood* upon a low lilac bush in front of my observing station. How it gleamed and changed in the reflected sunlight from across the street. It actually moved from branch to branch, and before I could realize in its semblance the figure of a bird, it had left a crimson, oleaginous impress upon my startled eyes. I had not, at that time had an introduction to or acquaintance with the Purple Finch, but as I watched his graceful movements, and listened to his low plaint, comfort and peace returned to my heart; and, as my Mother advanced into the room and enjoyed with me this natural beauty, I felt how sweet it was to be forgiven. This bird, her pardon, the homeless Frenchman, and sometimes the Calvary scene itself, have seemed to me synonymous; and, like well forged links in the petty changing chain of my youthful years, have bound the vision securely in the innermost chambers of my heart.

We may search far and wide for a created thing combining more sweetness of voice, beauty of form and feather, profuseness of love and gallantry and general good behavior and kindly disposition, than the male of the species under consideration. I have carefully noted him in early spring as he made his vows and promises to his homely clad but yielding and confiding mate.

For several successive seasons, a pair of these birds have had a favorite trysting place upon the well shingled roof of a neighbor's dwelling. When this little couple meet, the lordly one assumes all the consequence and importance of a turkey cock. His amply emarginate tail is fully spread, and his sheltering wings droop low and quiver with muffled vibrations upon the shingles, as he approaches near or retreats from his companion. During these ceremonious advances, he never once lowers his crimson crest or ceases to utter his coaxing notes of love.

Why he should have been christened with the " Purple " surpasses my comprehension. Mayhap some sprite upon whom the office fell, had, after indulging too freely in the nectar of a purple bowl, become color blind and knighted him with the tint of his favorite chalice; and, having no knowledge of color save as a name, bestowed upon this bird the

royal tint and tinted title. In its proper place, the reader will find a
full description of the livery of this handsome species.

Of late years, I have noted a seeming increase in the numbers of
these birds in this section, although I cannot recall a season in which
they have been uncommon. It is probably owing to this fact that they
have excited so much of curiosity and interest in the minds and eyes of
otherwise unobservant individuals. Their breeding localities also, as a
natural consequence seem to be tending toward those sections in which
their numbers are enlarging. Many of our birds once rare are becoming quite common, and *vice versa ;* whether this be owing to certain
physical changes in nature, to the advance of civilization and the artificial
work of man, or to some unknown cause, must for the present remain
to us a mystery. That this fact, regarding the species under consideration, has been noted by others than myself, there is abundant evidence to substantiate. Mr. R. F. Pearsall, of New York, in a communication to the Bulletin of the Nuttall Ornithological Club, for April
1879, says : "It has been a matter of remark that several of our once
rare birds have increased in numbers within a few years, and I think in
no case is this so apparent as in that of a Purple Finch. At the same
time its distribution extends over a much larger range. It was formerly
considered a strictly Northern migrant, but has recently become resident in Massachusetts, where it breeds quite plentifully in certain
sections."

The species, too, seems gradually but surely to be extending itself
each successive season into southerly localities ; the same writer continues : "Among some notes taken at Bayside, L. I., I find under date
of April 21, of this year. 'Saw a Purple Finch (male) in full song and
plumage and apparently resident.' In the early part of June I visited

the same locality and again saw both male and female. Feeling sure they must have nested there, after diligent search I discovered the nest, located, as usual, some forty feet from the ground, near the top of a large spruce tree, and contained only two eggs, well advanced in incubation." The writer quoted, surmises that this late domestic arrangement was probably owing to the first nest of the pair having been destroyed ; or that, as the actions of the birds seemed to indicate, they felt out of their latitude, at that, for the species, extreme southern point.

There seems to me, to be good reason for anticipating the capture of this bird, ere long, in Central New York during the Winter. Already it has become a Winter resident in the Hudson River Valley, where, in certain localities, it is at that season of the year comparatively abundant.

Mr. Edgar A. Mearns, of Highland Falls, N. Y., in his "Notes on some of the less hardy Winter Residents of the Hudson River Valley," in the organ of the Nuttall Club for January, 1879, says : "These beautiful birds and sweet songsters are regular Winter residents. In Winter there is always a preponderance of females. * * * *." Even the females are heard singing during the coldest weather ; this is of common occurrence in early Winter. They are gregarious, often assembling in very large flocks. On such occasions they are quite wild, and, on being approached, all rise at once on the wing, with a loud, rushing noise, accompanied by certain peculiar wild notes, which produces quite a startling effect. They feed upon seeds, chiefly those of the iron-wood (*Ostrya virginica*), and red cedar berries. "

Dr. Merriam, in his Review of the Birds of Connecticut, gives the species residence and records their abundant breeding. He further remarks, "Mr. Grinnell informs me that he has taken it during every

month of the year. Dr. Wood, of East Windsor Hill, tells me that
they were almost unknown here twenty years ago, and have gradually
become common since. I am likewise informed by Mr. Clark, of Say-
brook, that the bird has only recently become a common species in that
vicinity." In view of the increasing abundance, in various localities, and
the appearance in other sections of this bird, where it has been before
unknown or rarely met with, I shall not be greatly surprised to
either take or hear of its being taken, some mild and open winter in our
own latitude, as I have once already intimated.

Forty years since, the Purple Finch bred but sparingly in the northern sections of this state, as is evidenced by the writings of both Audubon and De Kay. Susan Fenimore Cooper, in her *Rural Hours*, under
date of Thursday, May 11th, (1848) says: "A large party of Purple
Finches also on the lawn ; this handsome bird comes from the far north
at the approach of severe weather, and winters in different parts of the
Union, according to the character of the season ; usually remaining
about Philadelphia and New York until the middle of May. Some few,
however, are known to pass the summer in our northern counties ; and
we find that a certain number also remain about our own lake, (Otsego)
having frequently met them in the woods, and occasionally observed
them about the village gardens in June and July. * * * * They
feed in spring upon the blossoms of flowering trees ; but this afternoon
they were eating the seeds out of decayed apples scattered about the
orchards." Referring to the nests about the door-yards and streets of
the village,—presumably Cooperstown,—the same author, writing February 22nd, 1849, says: "This last summer it looked very much as
though we had also Purple Finches in the village ; no nest was found,
but the birds were repeatedly seen on the garden fences, near the same
spot, at a time when they must have had young."

For a number of years past, this bird has bred regularly in the vicinity of Auburn; and to-day, the young collector can, without much trouble succeed in securing a nest of the species whensoever he may desire.

THE NEST AND EGGS.

I have before me, and through the kindness of my brother Sam. F. Rathbun, a nest of this Finch, which presents a very fair example of bird architecture in general. The structure measures in external diameter from three and three-fourths to four inches; in internal diameter two and one-half inches; and in depth, over all, one and seven-eighths inches. It is composed externally of the fine terminal twigs of various trees or bushes, among which, those of the spruce predominate. Intermingled with these, is an occasional half opened bud of the spruce, with now and then a small lanceolate leaf of some shrub or other. Succeeding this material, interiorly, and neatly arranged by the felting wings, the busy mandibles and dexterous feet of the builder, is a mixture of horse hairs—black, white and bay—together with shining shreds of the bark of dead twigs, and weed stalks. The lining of this nest is of wool principally, having a staple of some two and one-fourth inches long; this is mixed throughout with a few downy body-feathers and presents a comfortable concavity, almost perfectly semi-spherical of one and one-fourth inches in depth.

In such a tiny house, a veritable snug harbor of safety, my lady usually deposits four or five pledges to her faithful lord, each measuring in its normal condition, about three-fourths of an inch in length and nine-sixteenths of an inch in short diameter. These measurements I have taken from a specimen before me. Dr. Coues gives their average dimensions as 0.80 inches by 0.60 : but says, however, that he has seen specimens, from 0.85 by 0.67, (abnormally elongate) to 0.75 by 0.56. The color of the eggs is a clear greenish and light blue, irregularly dotted, chiefly at the largest end, with a very dark (almost black,) brown. The whole egg is in some instances, also very finely sprinkled with irregular markings of a washy purple. Occasionally the cow-bird leaves to the fostering care of the legitimists an offering of her own, and my brother has in his collection a set of six eggs, one of which is of this intrusive character.

Frequently the nest of the Purple Finch is built in an apple, maple or other tree. I have noted it in the maple, and the nest containing the set of six already mentioned, was taken from an apple tree, but a short distance from the ground. Usually, however, and with seeming preference, the nest of this bird is built some forty feet from the earth in some coniferous tree, generally the spruce ; where, lulled and soothed by the soughing breeze as it plays at hide and seek about her swaying dwelling, with the crimson banner of her loyal knight flaunting in and out amid the sombre green, and his musical tongue cheering her tedium, our lady fair broods from twelve to fourteen days ere she awakens to the full cares and joys of motherhood. During this period, our patient fair one in her frowsy state, is the recipient of the unremitting and watchful care of her mate, who yields to her his mellowest notes and plumpest captures, who gladdens her heart by his cheery disposition, and who dazes her little eyes with his brilliant plumes. Full of love, and full of anxious care she waits for Nature's pulses to perfect the little caskets committed to her charge ; which, being accomplished, she joins issue with her lord in search for tender food and in sentry duties about their joint and cozy cradle.

THE PLUMAGE, ETC.

The Purple Finch is from five and one-half to six inches in length, and his wings, when fully extended, measure from tip to tip, nine inches. His bill is of a brownish black above, but lighter beneath. The back is streaked with brown, while the wings and tail are of a deeper tint of the same color, the feathers being tipped and edged with reddish. A narrow band of cream color passes across the forehead ; all the rest of the body is of a rich, oleaginous crimson, passing into a rose color upon the belly, and fading into a dusky white at the vent.

The livery of the female differs very materially from that of the male, just described. The crimson tints of the latter are replaced in the former by a brownish-olive color, streaked with dusky and brownish. Her breast is of an ashy white or dusky, streaked with cuneiform spots of brown, while above and below the eye, appears a rather broad line of lighter colored plumage approaching to white. The plumage of the young birds of both sexes corresponds very closely to that of the mature female, save that in young males the rump and chin are of a bright olive yellow, which prevails also to some extent on the wing coverts, while the feathers of the tail are edged externally with olive.

Mr. William Brewster gives the *first plumage* of the female as follows : " Above dark brown, shading to lighter on the rump, each feather edged with light reddish-brown. The forehead and supra-loral line streaked with grayish. Under parts dull white, thickly streaked everywhere, except on crissum and anal region with very dark brown. From a specimen in my collection taken at Cambridge, (Mass.), July 9, 1873. Although this bird is in strictly first plumage, it differs scarcely appreciably in coloring from Autumnal specimens."—*Bulletin N. O. C.* Vol. III, p. 116.

We occasionally, but very rarely, find record of an albinistic tendency in the species. Mr. Ruthven Deane records the following : " Mr. George A. Boardman has in his fine collection, in which so many albino birds are represented, a fine pure white Purple Finch ; and through the kindness of Mr. H. Herrick of New York, I have in my collection a dull cream colored bird of this species which he shot at Umbagog Lake, Me., some years ago."—*Ibid.,* Vol. IV. : p. 28.

The plumage of both sexes of this Finch is generally close and compact, beautiful in its blendings, and colored with firmness and harmony. The neck is rather short, head somewhat large, and the body plump and symmetrical. The bill is conical, slightly convex above and pointed ; tail, forked ; wings, of moderate length, third and fourth primaries longest.

Audubon says his tongue is five-twelfths of an inch long, sagittate, papillate, channeled above, etc., etc., but what a tongue ! As this musician balances himself at an aspiring height in the air, upon his passion laden wings, we may think and write of the poetry of sound, of the soothing cadences of " complaining brooks," or the chiming harmonies of nature ; but the only thought of the ear that listens to the matin roundelay of this bird, is, that they are sweet, *so sweet* and truthful ; and of the heart they touch, that they are *full,* FULL of love and purity.

They fall upon the head of the toiler under the arch of heaven like a benison, and on the studious and reflective mind like a benedictive gift from the weaving and wonder-working fingers of Nature, and are fit pæans for the heralding of her prolific feet.

AUDUBON ON THE PURPLE FINCH.

ERYTHROSPIZA PURPUREA, GMEL.

Plate cxcvi.—MALE AND FEMALE.

" From the beginning of November until April, flocks of the Purple
Finch, consisting of from six to twenty individuals, are seen throughout
the whole of Louisiana and the adjoining States. They fly compactly,
with an undulating motion, similar to that of the Common Green Finch
of Europe. They alight all at once, and after a moment of rest, and as
if frightened, all take to wing again, make a circuit of no great extent,
and return to the tree from which they had thus started, or settle upon
one near it. Immediately after this, every individual is seen making its
way toward the extremities of the branches, husking the buds with
tact, and eating their internal portion. In doing this, they hang like so
many Titmice, or stretch out their necks to reach the buds below. Although they are quite friendly among themselves during their flight,
or while sitting without looking after food, yet, when they are feeding,
the moment one goes near another, it is strenuously warned to keep off
by certain unequivocal marks of displeasure, such as the erection of the
feathers of the head and the opening of the mouth. Should this intimation be disregarded, the stronger or more daring of the two drives
off the other to a different part of the tree. They feed in this manner
principally in the morning, and afterwards retire to the interior of the
woods. Towards sunset they reappear, fly about the skirts of the fields
and along the woods, until, having made choice of a tree, they alight,
and, as soon as each bird has chosen a situation, stand still, look about
them, plume themselves, and make short sallies after flies and insects,

but without interfering with each other. They frequently utter a single rather mellow *clink*, and are seen occupied in this manner until near sunset, when they again fly off to the interior of the forest. I one night surprised a party of them roosting in a small holly tree, as I happened to be brushing by it. In their consternation they suddenly started all together, and in the same direction, when, not knowing what birds they were, I shot at them and brought down two.

" It is remarkable that, at this season, males in full beauty of plumage are as numerous as during the summer months in far more northern parts, where they breed ; and you may see different gradations of plumage, from the dingy greenish-brown of the female and young to the richest tints of the oldest and handsomest male : while along with these there are others, which by my habit of examining birds, I knew to be old, and which are of a yellowish-green, neither the color of the young males, nor that of the females, but a mixture of all.

" The song of the Purple Finch is sweet and continued, and I have enjoyed it much during the spring and summer months, in the mountainous parts of Pennsylvania, where it occasionally breeds, particularly about the Great Pine Forest, where, although I did not find any nests, I saw pairs of these birds flying about and feeding their young which could not have been many days out, and were not fully fledged. The food which they carried to their young consisted of insects, small berries and the juicy part of the cones of the spruce pine.

" They frequently associate with the Common Cross-bills, feeding on the same trees, and like them are at times fond of alighting against the mud used for closing the log-houses. They are seldom seen upon the ground, although their motions there are by no means embarrassed. They are considered as destructive birds by some farmers, who accuse them of committing great depredations on the blossoms of their fruit-trees. I never observed this in Louisiana, where they remain long af-ter the peach and pear trees are in full bloom. I have eaten many of them, and consider their flesh equal to that of any other small bird, ex-cepting the Rice Bunting.

"This species was seen by Dr. Richardson on the banks of the Saskatchewan River only, where it feeds on willow buds. It arrives there in May, and resides during the summer. The eggs have been procured in the State of Massachusetts by my friend Dr. T. M. Brewer. They measure seven-eighths and a quarter in length, four-eighths and a half in breadth, and are thus of an elongated form, rather pointed. Their ground color is a bright emerald-green, sparingly marked with dots and a few streaks of black, accumulated near the apex, and some large marks of dull purple here and there over the whole surface.

*   *   *   *   *   *   *   *   *   *   *

"I have found this species from Labrador to Texas. Mr. Nuttall and Mr. Townsend met with it on the Columbia River, and all the way to St. Louis. In South Carolina, where it appears only during severe winters, it feeds on the berries of the Virginia juniper, commonly called the red cedar; and when the berries fall to the ground, it alights and secures them. Dr. Bachman has kept it in aviaries, where it became very fat, silent, and only uttered its usual simple feeble note. After moulting, the males assumed the plumage of the females. The next spring a very slight appearance of red was seen, but they never recovered their original brilliancy, and it was difficult to distinguish the sexes. It breeds sparingly in the northern parts of the State of New York. In June 1837, I met three pairs, within a few miles of Waterford, that evidently had nests in the neighborhood."

RANGE OF THE PURPLE FINCH.

Regarding the range of this bird I can do no better than refer
to Dr. Coues. This writer, in *Birds of the North-west*, gives the species
habitation from the Atlantic to the Pacific, excepting, probably, the Southern Rocky Mountain region, where it is replaced by Cassin's Purple
Finch and the Crimson-fronted Finch or Burion. North to the Saskatchewan and Labrador—Wintering in the Southern States.

FIG. OF THE PURPLE FINCH.

BRIGHT FEATHERS

OR

Some North American

BIRDS of BEAUTY.

By Frank R. Rathbun.

*Illustrated with Drawings made from Nature, and carefully
Colored by Hand.*

AUBURN, N. Y.
PUBLISHED BY THE AUTHOR.
1881.

GONIAPHEA LUDOVICIANA, (Linn.) Bowd.

THE ROSE BREASTED GROSBEAK. ♂·MALE. ♀·FEMALE. Life Size.

DRAWN FROM NATURE. BY FRANK R. RATHBUN.

BRIGHT FEATHERS.

PLATE II.

THE ROSE-BREASTED GROSBEAK.

GONIAPHEA LUDOVICIANA, (LINN.) BOWD.

PLATE II.—MALE AND FEMALE.

F the various feathered forms that illuminate the verdure of our woods and groves, none are more noticeable for their plumage colorations than the subject of these pages. Striking, from their very simplicity and the well defined contrasts of black, white and rose color, the markings of the Rose-Breasted Grosbeak are sure to challenge the notice of the most indifferent observer.

ORIGINAL INITIAL.

Referring to my field book, I find under date of May 4th, 1880, a note of the first appearance of this bird in Auburn for the season. The scenic effects under which it was welcomed, were of the most auspicious nature, and to my mind boded a forecast of long and sunny, leafy days. The year was shaking itself awake from a lethargic sleep, bursting the bonds of a prolonged slumber; and the thermometer placed in the shadow of a doorway, registered 88° Fahr.

Sauntering through South Street in the forenoon of the day mentioned, I noted the evidences about me of the advent of the season. The swelling embryos of silent life seemed ready to burst into being at any moment, having only to await the vivifying touch of an invisible finger more cunning and curious in its ways than that of any mechanical artificer. The boughs of the graceful elms, laden with their tiny plumes, seemed to nod me a good morrow as I passed, and upon the fences, coming from whence I know not, the spiders were seen sunning their sprawling forms. The tender blades of golden hued grasses were gently crowding aside the dead leaves of the preceding autumn, like true lollards of a murmuring hour. Spring had cast her hazy banner athwart a cloudless sky, while up from the South came a warm breath, laden with langour and laziness, which found voice enough to murmur, "The Birds are Coming! The Birds are Coming!" Over the distant lake there seemed to rest an undefined, but nevertheless distinguishable humidity that seemed only too willing to lend its vivifying influence to the scene, and turned and swayed in the atmospheric haze like a true toiler in the performance of an allotted task.

Suddenly, and without warning, there burst upon my ears a note, clear, incisive and well defined, full of nuptial pride and melody, each repetition sweeter and more thrilling than its precursor, which I recognized at once as emanating from no other source than the throat of the Rose-Breasted Grosbeak. It was no difficult task for me to quickly discover and identify the minstrels of this pleasing lay, for, rollicking in almost every variety of attitude amid the twigs of a stately elm, my eyes were gladdened by the sight of three males of the species, whose cenobic garb and rosy shield displayed in full perfection, and pro-

claiming their full maturity, told only too plainly of their name and
lineage.

On the 14th day of the same month, I was sent for by one of my
lady acquaintances on Clark Street of this city, to identify a very novel
bird capture.   I immediately proceeded to her home, and there found
confined in a large cage, which was suspended in the grateful shade of
an eastern porch, (the day being exceedingly warm for the season,) a
beautiful and perfectly plumed Rose-Breasted Grosbeak.   The mien
of the handsome captive was bold and fearless, evincing no apparent
care or trepidation in its continement.   I was informed that it had been
captured in the green-house of the premises, through the open door of
which it had entered, doubtless attracted by the tropic luxuriance therein.   The ambitious florist of the establishment, after closing the doors
of the greenery, had brought to bear upon his strange guest a stream
of water from the garden hose and had thus stunned and bewildered the
unfortunate bird to such an extent, as to render his capture an easy
task.   Cool and humid as his welcome had been in the greenery, he
evinced comparative contentment in his wiry prison, and, after pluming
his feathery coat, fed fearlessly of the seeds and crumbs placed at his
disposal.   The kind and womanly heart of his mistress, however, could
not reconcile his confinement to her generous instincts, and he was sub-
sequently released from durance, whence he sped easterly in glad-
ness of heart and strength of wing.

There can be no doubt but that this bird would bear confinement
well and would certainly make a pleasing and most desirable acquisition
to any aviary.   Mr. C. H. Merriam in his *Birds of Connecticut*, says:
" I am informed by Prof. G. Browne Goode, of Middletown, Conn., that

he knew an individual of this species to live eighteen years in confinement." —*Birds of Connecticut*, p. 43.

Although the Rose-Breasted Grosbeak is not abundant enough in this section to command the universal acquaintance and attention of the majority of people it is nevertheless not an uncommon summer resident; and under whatever circumstances, is sure to enlist the peculiar notice of the fortunate observer. It arrives here about the middle of May, and leaves for the South in the latter part of September.

Reference has already been made to the pleasing note of this species, and I cannot refrain from quoting Dr. Elliott Coues in this particular. He says: "I have nowhere found this beautiful bird more abundant than along the Red River of the North, and there may be no locality where its nidification and breeding habits can be studied to greater advantage. On entering the belt of noble timber that borders the river, in June, we are almost sure to be saluted with the rich, rolling song, of the Rose-Breasted male."—*Birds of the North-West*, p. 166–167.

THE NEST AND EGGS.

The same author continues. "As we penetrate into the deeper recesses, pressing through the stubborn luxuriance of vegetation into the little shady glades that the bird loves so well, we may catch a glimpse of the shy and retiring female, darting into concealment, disturbed by our approach. She is almost sure to be followed by her ardent spouse, solicitous for her safety, bent on reassuring her by his presence and caresses. Sometime during this month, as we enter a grove of saplings, and glance carefully overhead, we may see the nest, placed but a few feet from the ground, in the fork of a limb. The female, alarmed, will flutter away stealthily, and we may not catch another glimpse of her,

nor of her mate even, though we hear them both anxiously consulting
together at a little distance.   The nest is not such an elegant affair as
might be desired ;   it is, in fact, bulky and rude, if not actually sloven-
ly.   It is formed entirely of the long, slender, tortuous stems of woody
climbers, and similar stout rootlets ; the base and outer walls being very
loosely interlaced, the inner more compactly woven, with a tolerably firm
brim of circularly disposed fibres.   Sometimes there is a little horse
hair lining, oftener not.   A very complete nest before me is difficult
to measure from its loose outward construction, but may be called six
inches across outside, by four deep ; the cavity three inches wide, by one
and a half deep.   The nest contained three eggs, which I think is the
usual number in this latitude ; four I have only found once.   The
eggs are usually rather elongate, but obtuse at the smaller end.   Different specimens measure, 1.00 by 0.75, 1.08 by 0.70, 1.03 by 0.75, 1.02
by 0.72 ; 0.96 by 0.76 ; by which dimensions the variation in shape is
denoted.   The average is about that of the first measurement given.
They are of a light and rather pale green color, profusely speckled with dull
reddish-brown, usually in small and also rather diffuse pattern, but sometimes quite sharpely marked ; the sharper markings are usually the small-
est.   There is sometimes much confluence, or at least aggregation, about
the greater end, but the whole surface is always marked.   Most of these
eggs were taken in the latter part of June, some by the end of the
month ; in all, incubation was in progress."—*Ibid.*

My brother, Sam., has handed me the following measurements of
two sets of eggs in his collection.   Set of 3. No. 1, .99 by .74 in. ; No.
2, 1.00 by .75 ; No. 3, 1.00 by .76.   Set of 4. No. 1, 1.10 by .75 in ;
No. 2, 1.08 by 75 ; No. 3, 1.05 by .73 ; No. 4, 1.00 by 74

THE PLUMAGE, ETC.

The coloration of the head and upper parts of the male of the species, is glossy black; the breast and under wing coverts, rose color, which in some instances descends down the centre of the breast as far as the vent. The posterior portion of the back, or rump, the breast and belly, (where not tinted with rose,) white. Two bands on the wings, base of the quills, tips of the secondaries, and the terminal half of the three inner webs of the outer tail feathers, also white. In some instances the rump is spotted with black. The plumage is soft and elastic and the colorations in the adult bird, well, but not too hardly defined, excepting perhaps the line bounding the black of the throat, and the rose color upon the breast. The bill is almost white, short, sharp and robust, the outer lines of both mandibles being a little convex, and the inner lines deflected at the base, thence straight to the end. Legs of medium length and stout, with a few longish anterior scutella upon the tarsus; posteriorly the latter is sharp. Toes, well divided and scutellate above, the lateral ones being nearly equal in length. Claws rather slender, acute and arched. Tail slightly forked, somewhat long, and consisting of twelve feathers. Iris, hazel; feet and legs, bluish; mien, bold and free; body, well developed and rounded. The male does not attain generally to full maturity of plumage, until about the third year. Length, $7\frac{3}{4}$ inches; alar extent, 13; bill along the back $\frac{9}{10}$, along the edge $\frac{7}{10}$; tarsus $\frac{10}{12}$. (Audubon). Length 7.5.—8.5, alar extent, 12.5—13.0 (DeKay.) Specimen before me, probably a bird of the second year, $7\frac{1}{2}$ inches long, and $12\frac{5}{8}$ inches alar extent.

The female differs very materially from the male in color markings, as will be seen on referring to the plate. Beneath, it is of a slight reddish or yellowish white marked with spots of brown. Above olivaceous brown with darker markings. Under wing coverts, saffron yellow. Central band on the head, and on each side of darkish brown; above the eye, a white one. A brown stripe from the bill to the eye and beyond; under this a stripe of whitish. Quills and tail, brown. Bill brown above, lighter beneath. Iris and feet as in the male. According to Audubon, the young male after the first moult, resembles the female, excepting that it then begins to show the rosy tints both on the breast and under wing coverts. In the first plumage the young resemble the female.

My friend, Mr. Edgar A. Mearns, of Highland Falls, N. Y., has very kindly furnished me with the following notes relating to the habits of this beautiful bird:

MEARNS ON THE ROSE-BREASTED GROSBEAK.

Notes on the Habits of the Rose-Breasted Grosbeak. (Zamelodia Ludoviciana).

In the Hudson Valley, this handsome Grosbeak is a summer resident, and breeds plentifully in certain districts. It comes from the South early in May, and continues with us until the end of September. On its first arrival, it is usually seen in the woodlands, haunting the lower branches of the trees, where its bright colors at once attract attention. During the past spring (1880), the pioneer Grosbeak was a beautiful male seen perched in a tree-top on Cro'-Nest Mountain, near West Point, on the 4th of May. Soon after its arrival, the Rose-Breasted Grosbeak appears about our houses, and, possibly does some damage to the fruit crops, by eating the blossoms in the orchards; it is

especially fond of those of the cherry, and the rapidity with which it
dispatches them is quite marvelous.   It is not, however, regarded with
disfavor; perhaps because its beauty and delicious song disarm suspi-
cion or allay punishment.   It also resorts to the fields, newly sown
with the cereal grains, when such are convenient to its woodland retreats, and there will collect in large flocks, accompanied by the Purple Finch, Indigo Bird, and a host of other fringilline depredators; owing to its black and white colors, these flocks are frequently mistaken
by the farmers for flocks of Bobolinks.   When frightened, the whole
flock flies into the nearest trees, with a loud whirring of wings, and uttering its characteristic sharp note.   A giant white oak stands in the
line of a stone wall, in sight of my house; when the surrounding fields
are sown with grain, I have there observed the Grosbeak in larger
flocks than elsewhere; when alarmed, they invariably fly directly into
the big oak tree; to stand near it when the birds take to flight, is to
witness a pleasing sight, for a vivid crescent of carmine glows upon the
snowy bosom of each male, and his mate exhibits to best advantage the
beautiful yellow lining of its wings, which, in the male, is bright crimson,
and its whirring flight and singular note are quite striking.

The Rose-Breasted Grosbeak evinces a preference for highlands,
during summer.   The swampy margin of some mountain brook or
pond is the most likely place to hunt for its nest, especially should the
male bird be heard singing near by.   The nest is built in the thickest
part of some low tree; it is coarsely constructed of leaves, sticks, stems
of plants, grass, bits of bark and similar materials found about its
neighborhood; near the nest, the male may be heard singing much of
the time during the incubating period.   The eggs are very similar to
those of the Scarlet Tanager, (*Pyranga rubra*), and are green, blotched

with reddish-brown, averaging .98x.70 of an inch. This Grosbeak possesses a fine song, at times quite loud, which it warbles constantly during spring, and through the nestling season.

EDGAR A. MEARNS.

Highland Falls, N. Y., Aug. 19, 1880.

THE GROSBEAK vs. THE POTATO BUG.

Althought this beautiful bird has been severely chastened by some writers for his depredations among the buds and blossoms of our fruit trees and orchards, he seems to be developing a commendable offset thereto in his partiality for the potato beetle. I have carefully collated the following notes relating to this fact, which will explain themselves without further prefatory :

" ROSE-BREASTED GROSBEAK AND COLORADO POTATO BEETLE.—Regarding this useful and pleasing bird, the following appeared in the *New York Weekly Tribune*, of February 11, 1880, to wit : " Prof. C. E. Bessey, of the Iowa Agricultural College, several years ago observed the Rose-Breasted Grosbeak's habit of feeding on the Colorado potato beetle. ° °. Its useful propensity was again remarked during the past year by a correspondent of *Forest and Stream*, at Coralville, Iowa, and by another at Ames in the same State." In a small aviary which I keep for better observing the habits of our native and several foreign song birds, the same preference in the selection of food was noticed by myself. September 18, 1879, I found in the flower bed of my yard, a potato beetle (*Doryphora decem-lineata*), which I intended to give to my Cardinal Grosbeak (*Cardinalis virginianus*). After placing it in the cage, it was with difficulty that I prevented the Rose-Breasted Grosbeak (*Hydemeles Indoviciana*) from seizing it at once. Knowing the potato beetle to be poisonous, at least to the genus *Homo*, I did not care to try experiments with the tame Rose-Breasted Grosbeak, my sweetest songster in the aviary. I had often before noticed, that the Cardinal Grosbeak had a fondness for beetles, and naturally supposed that he was better qualified to judge of the wholesomeness of the food offered."

"The latter bird watched the beetle very attentively as it crept over the floor, but seemed to be in no hurry to capture it as he would other beetles. Finally he took hold of and crushed it between his mandibles. He tried to swallow it, but not finding the taste very appetizing, he gladly yielded up the unsavory morsel to the repeated snatchings of the Rose-Breasted Grosbeak. After crushing it into a shapeless mass, which occupied but a few seconds, he carefully stowed it away. I thought he seemed not to enjoy the taste very much, as he gravely shook his handsome head as if in a doubtful mood. Wiping off his bill, he immediately afterwards proceeded to eat a quantity of cultivated portulaca that I offered as a precautionary measure. Portulaca stems and leaves contain much mucilage, which I thought would be an antidote to the possible acrid quality of the juice of *Doryphora decem-lineata*. The usual result of poisoning from handling these crushed beetles, as well as from inhaling the fumes arising from vessels in which *D. decem-lineata* have been scalded, has been likened to serpent and scorpion-poisoning. Where death followed, the blood would become disorganized the same as from septæmia. However, in the case of the birds no evil effects were noticed. In mankind, idiosyncrasy favors the absorption of the volatile doryphora poison."

      ❋   ❋   ❋   ❋   ❋   ❋   ❋   ❋   ❋   ❋

            "RICHARD E. KUNZE, New York."

—*American Naturalist*, Vol. XIV., pp. 521–522.

"THE POTATO BUG BIRD.—*Coralville, Iowa.*—When the Colorado beetle created such sad havoc among the potato vines here in the West, the potatoes were uninjured in this vicinity for several seasons, owing to the kind offices of some beautiful birds, familiarly termed as above. Though they had previously visited and "bugged" the potatoes on some farms located in the timber five miles distant, for four or five successive seasons, it was not until the summer of 1877 that they turned their attention to this place. In the summer referred to, the bugs appeared as usual in great numbers and began working on the vines, and would, if unmolested, have destroyed them in a little while. One morning my attention was drawn to some strange, handsome birds that were sitting on the garden fence, surveying our potato grounds. I watched them closely for some time. There appeared to be but a single family of them ; the two parent birds, and some half a dozen young ones. The latter were arranged in a row on the topmost board, awaiting their turns to be fed ; and it kept the old ones very busy to attend to them. They would flit down in among the vines and seize an unlucky bug, and carry it to a little open mouth, again, and again, till I wearied of watching them. I knew at once that this must be the potato bug bird, of which I had so often heard. This was repeated day after day, till our garden was cleared of the bugs. They then helped themselves to a few—a very few—peas (for dessert, I suppose), and then commenced work on the potatoes in an adjoining field. In size and shape this bird resembles the orchard oriole ; is black and white in color, with a handsome red spot on its breast—reminding one of the prevailing hue of the potato beetle—and black and white bands across the wings." W.

"Perhaps the bird which does this good service is the Owhee Bunting (*Pipilo erythrophthalmus*). We should like to hear more about it."
—*Forest and Stream*, Vol. XIII, p. 827, Nov. 20, 1879.

"THE POTATO BUG BIRD IDENTIFIED,—I wish to say something confirmatory of the information given by "W" from Coralville, Iowa, in *Forest and Stream* of 20th ult. There are undoubtedly many—farmers and others—whose experience would be much more valuable than mine, but in the vicinity of St. Paul, Minnesota, I have often seen the birds engaged in feeding themselves and their young, just as described by "W," yet I have not seen them so acting at any great distance from groves or timber. I do not think "W" has overstated any of the good or pleasing qualities of the bird. It is not, however, the Bunting suggested by the editor, and known as owhee, towhee, chowee, che-wing, joe-ree, etc., but is the Rose-Breasted Grosbeak, ("May his tribe increase,") and is the only bird of which I have any knowledge or information whose first choice in matter of diet is a fresh potato bug. Of such is its favorite meal, and it does not often call for dessert or condiments when its preferred game is in season. But few farmers and gardeners make themselves acquainted with their best friends, or the bird with the rose-tinted bosom, would have his name inscribed well up towards the head of the list ; so high, indeed, that our old friend, Robin Red-Breast, would have to put himself upon his good behavior, in order to retain his standing."   M.

PEMBINA, D. T., Nov. 30th.

"MINNESOTA—*Minneapolis, Nov. 24th.*—In your issue of November 20th, appeared a communication from a correspondent in Iowa, in regard to a certain bird eating the potato bug, or Colorado beetle. The bird in question is undoubtedly the Rose-Breasted Grosbeak (*Goniaphea ludoviciana*). The description given by your correspondent applies to this species, and the habit to which he calls attention has been noted in this locality by the writer and others for several years past. I have seen a number of these handsome birds in and about a single potato patch at once, and during the period, when the bugs are at work, the Grosbeaks are regular frequenters of the potato field."

" But the bugs are legion, and the Grosbeak's task becomes a mighty one. Yet he seems willing to do his part, and should be awarded the hearty thanks and firm friendship of the farmers he befriends."

"T. S. ROBERTS."

"ILLINOIS—*Normal, Nov. 22d.*—The bird referred to in the interesting note of your correspondent at Coralville, Iowa, is doubtless our Rose-Breasted Grosbeak (*Guiraca ludoviciana*), the only bird which, in the critical examination of the contents of several hundred stomachs of birds, I have found to make a practice of eating the Colorado potato beetle. In fact, with the exception of a single specimen found in the stomach of a Robin, I have not found this beetle in the stomach of any other bird."

" This beautiful bird, as attractive for its clear, rich, and striking song as for its showy plumage, is increasing rapidly in numbers in this vicinity ; but, hitherto, I have credited it with good intentions respecting the potato pest rather than with any effective service. I am glad to learn that it is capable of becoming a real protection to the farmer's crop. Unfortunately, it offers temptations to the taxidermists—the pot hunters of ornithology—and is becoming too common in sets of fancy skins for the good of agriculture." "S. A. FORBES."

—*State Lab. Nat. Hist.*, Normal, Ill. *Ibid.* Vol. XIII, p. 907, Dec. 1879.

"THE POTATO BUG BIRD AGAIN.— *Ames, Iowa, Nov. 26th, 1879.*— In your issue of Nov. 20th, I notice a note from "W." of Coralville, Iowa, in which he speaks of a bird which feeds upon the Colorado potato beetle. This bird is the Rose-Breasted Grosbeak—*Goniaphea ludoviciana,* Bowditch. The first who noticed its habit of feeding upon the potato beetle was, as far as I know, Professor Bessey, of the Iowa Agricultural College, who observed it several years ago. I confirmed the observation during the past summer, and am glad to get this additional proof from Coralville. As the potato beetle is a comparatively recent comer in the State, it is of course new food for the bird which seems to be just finding out this strange and abundant supply of provision. It is to be hoped that this will prove one of those natural enemies to the insect which we have been looking for, and which will restore the balance of nature which has been so sadly disturbed in the case of the potato beetle for the past ten years."

                                              "F. E. L. BEAL."

–*Ibid.* Vol. XIII, p. 1005. Jan. 22. 1880.

AUDUBON ON THE ROSE-BREASTED GROSBEAK.

COCCOBORUS LUDOVICIANUS, LINN.

Plate ccv. —MALE, FEMALE AND YOUNG.

"One year in the month of August, I was trudging along the shores of the Mohawk River, when night overtook me. Being little acquainted with that part of the country, I resolved to camp where I was ; the evening was calm and beautiful, the sky sparkled with stars, which were reflected by the smooth waters, and the deep shade of the rocks and trees of the opposite shore fell on the bosom of the stream, while gently from afar came on the ear, the muttering sound of the cataract.

My little fire was soon lighted under a rock, and, spreading out my scanty stock of provisions, I reclined on my grassy couch. As I looked around on the fading features of the beautiful landscape, my heart turned towards my distant home, where my friends were doubtless wishing me, as I wished them, a happy night and peaceful slumbers. Then were heard the barkings of the watch-dog, and I tapped my faithful companion to prevent his answering them. The thoughts of my worldly mission then came over my mind, and having thanked the Creator of all for his never-failing mercy, I closed my eyes, and was passing away into the world of dreaming existence, when suddenly there burst on my soul the serenade of the rose-breasted bird, so rich, so mellow, so loud in the stillness of the night, that sleep fled from my eyelids. Never did I enjoy music more ; it thrilled through my heart, and surrounded me with an atmosphere of bliss. One might easily have imagined that even the Owl, charmed by such delightful music, remained reverently silent. Long after the sounds ceased did I enjoy them, and when all had again become still, I stretched out my wearied limbs, and gave myself up to the luxury of repose. In the morning I awoke vigorous as ever, and prepared to continue my journey."

"I have frequently observed this beautiful species, early in the month of March, in the lower parts of Louisiana, making its way eastward ; and when residing at Henderson in Kentucky, and in Cincinnati in Ohio, I have noticed the same circumstance. At this early period, it passes at a considerable height in the air, and now and then alights on the tops of the tallest trees of the forest, as if to rest awhile. While on wing it utters a clear note, but when perched it remains silent, in an

upright and rather stiff attitude. It is then easily approached. I have
followed it in its migrations into Pennsylvania, New York, and other
Eastern States, through the British Provinces of New Brunswick and
Nova Scotia, as far as Newfoundland, where many breed, but I saw
none in Labrador. It is never seen in the maritime parts of Georgia,
or those of the Carolinas, but some have been procured in the mountainous portions of those States. I have found them rather plentiful
in the early part of May, along the steep banks of the Schuylkill river,
twenty or thirty miles from Philadelphia, and observed, that at that
season they fed mostly on the buds of the trees, their tender blossoms,
and upon insects, which they catch on wing, making short sallies for the
purpose. I saw several in the Great Pine Forest of Pennsylvania;
but they were more abundant in New York, especially along the banks
of the beautiful river called the Mohawk. They are equally abundant
along the shores of Lakes Ontario and Erie, although I believe that the
greater number go as far as New Brunswick to breed. While on an
excursion to the islands at the entrance of the Bay of Fundy, in the
beginning of May, my son shot several which were in full song. These
islands are about thirty miles distant from the main land."

" The most western place in which I found the nest of this species
was within a few miles of Cincinnati on the Ohio. It was placed in
the upright forks of a low bush, and differed so much in its composition
from those which I have seen in the Eastern States, that it greatly
resembled the nest of the Blue Grosbeak already described. The
young, three in number, were ready to fly. The parents fed them on
the soft grains of wheat which they procured in a neighboring field, and
often searched for insects in the crannies of the bark of trees, on which
they alighted sidewise, in the manner of Sparrows. This was in the

BRIGHT FEATHERS

OR

Some North American

BIRDS of BEAUTY.

By Frank R. Rathbun.

Illustrated with Drawings made from Nature, and carefully Colored by Hand.

AUBURN, N. Y.
Published by the Author.
1882.

end of July. Generally, however, the nest of the Rose-breasted Grosbeak is placed on the top branches of an alder bush, near water, and
usually on the borders of meadows or alluvial grounds. It is composed
of the dried twigs of trees, mixed with a few leaves and the bark of
vines, and is lined with fibrous roots and horse-hair. The eggs are
seldom more than four, and I believe only one brood is raised in the
season. Both sexes incubate. I have found the nest and eggs, on the
20th of May, on the borders of Cayuga Lake, in the State of New
York."

"The flight of the Rose-breasted Grosbeak is strong, even, and as
graceful as it is sustained. When travelling southward, at the approach
of autumn, or about the 1st of September, it passes high over the forest
trees, in the manner of the King-bird and the Robin, alighting toward
sunset on a tall tree, from which it in a few minutes dives into some
close thicket, where it remains during the night. The birds travel
singly at this season as well as during spring."

CHRYSOMITRIS TRISTRIS, (Linn.) Bonap .

THE AMERICAN GOLDFINCH. ♂·MALE.♀FEMALE.Life Size.

DRAWN FROM NATURE BY FRANK R.RATHBUN.

Bright Feathers .

plate III .

THE AMERICAN GOLDFINCH.

CHRYSOMITRIS TRISTRIS. (LINN.) BONAP.

PLATE III.—MALE AND FEMALE.

OST people have an intimate acquaintance with this innocent and happy little bird. Probably, no other single species has had so many local appellations bestowed upon it as the one of which we now attempt a simple history and description. Known in various localities as the Yellow-bird, he has also, from peculiarities of dress, habit and surroundings, been christened the Hemp-bird, Lettuce-bird, Thistle-bird, Thistle-finch, Salad-bird, Yellow-finch, Black-winged Yellow-bird and Goldfinch, besides other appellations spontaneously bestowed from love, and with careless, though in many cases with apt freedom by those who welcome his presence and cultivate an acquaintance with his sweet and winning ways.

ORIGINAL INITIAL.

The various names which I have cited, are, from the very apparent intuition of their bestowal, a fitting index to the reflective mind, of the habits and dress of the American Goldfinch. In some isolated sections of

Northern Vermont, I have heard him named the " Wild Canary-bird," from which one might infer that he was of near kinship to the familiar pet of our homes, and be not very far removed from fact so far as his size, form, and general habit is concerned. He is also known, in some parts of this country, as the Siskin ; a name derived from the Danish, *Sisgen* ; the Swiss, *Siska* ; the German, *Zeisig* ; which is bestowed in most cases by foreigners, and which refers him by analogy, to the Aber-devine or European Siskin which belongs to the same sub-species as his English congener, the Goldfinch.

Amid this redundant array of names however, the Common Yellow-bird or American Goldfinch pursues the cheery, even tenor of his way ; a pattern of modesty, sweet of voice, void of an excess of timidity or intrusiveness and very fond of the company of his own kith and kin. His satin like robe of golden complexion, renders him a conspicuous object on all occasions, and his modest familiarity, for once, has failed to breed the contempt of the well known proverb.

This bird is a free rover. In the larger cities as well as in the villages and hamlets, he may be found as free and unfettered by care as in the open country. Should a city dweller have the privilege of a small garden plot, and for his own edification and relish choose to cherish a crisp bed of lettuce for the compounding of his own salads, he may, should he allow a few roots to assert their privilige, and grow to maturity, be sure some bright afternoon as he goes out to collect and save the seed for another season, to find a bevy of these little birds ahead of him, industriously appropriating for their own food, the cherished germs of his choice variety. On such an occasion, should the feast prove ample and inviting, these birds may be approached quite near, apparently heedless or careless of the proprietor of the mimic glebe.

Should one stroll into the open air for health and relaxation, he will, especially if at the culminating period of vegetation, be sure to meet with whole troops of these finches clinging to the sturdy stalks, and flitting among the dusty, trailing gowns of the invincible and waving weeds. Especially does the Goldfinch seem to favor and love the well armored " Bull thistle " of the school boy ; *Cnicus lanceolatus* of systematists ; the emblem of " Bonnie Scotland ;" the sentinel knight of lordly mien and a thousand well tempered lances ; whose helm is *purpure*, and whose motto " *Nemo me impune lacessit*," is a pointed challenge to the rash one who would deflower him ; but whose bolls, when sere and brown with ripeness, are ruthlessly torn open by our yellow friend who regales himself upon their myriad seeds, the downy wings of which he scatters upon either side, to be dandled by the passing breeze.

The stroller, as he continues his easy tramp, now over a rustic bridge which spans a laughing shallow stream or a moist depression in the highway, or by the damp sedge of a leaning length of fence, as well as upon the very highway itself, is hailed by romping congregations of this bird and welcomed to the freedom of their haunts. In the latter part of August, when the conditions already described have culminated, he will not fail to notice that the major number of these happy forms seem to be of nearly one uniform tint or complexion. Now and then, it is true, a finch may be noted that still bears a brightly tinctured coat when seen from a short distance ; but which, upon closer inspection will reveal an undertone of tawny hue displacing the auriferous plumes of the vernal season, soon to eventuate in a uniformity of dress for both of the sexes. The fact is, that at this time, as a prominent writer upon such topics has well said, " The Goldfinches doff satin for linen," in which humble garb the Quaker complexioned band is just as happy, just as free, and just as unconcerned, careless and gay, as though the lords wore their topaz garniture and the ladies were the pleased and contented recipients of their courtly gallantries.

This plumose change of the birds has excited the discussion and close observation of many eminent naturalists. Mr. George Ord, writing in March 1828, with a view to ascertaining whether the opinion of Temminck, that some birds changed their plumage *twice* a year was founded in fact, says : "This change takes place, in some species, in summer ; in others in autumn. When the old feathers drop, their place is supplied by new ones, which, in some species, are of quite a different complexion from those that they succeed. But when, in the spring, a retrocession of color is found to have taken place, naturalists have concluded that these birds undergo a double moulting ; for in no other way could they account for the change of color, which has been supposed to be dependent upon a change of plumage. The species which are usually domesticated have been said to moult but once a year, because, not perceiving any material change in their garb, it is inferred that no change is necessary, and yet, if any notable mutation had obtained in any one of the domesticated species, it is probable it would be affirmed of that species that there was some physical necessity for this exception which did not hold of the rest."

"The intention of Nature in renewing the covering of birds appears to be a reinvigoration of those powers which are necessary to the propagation and conservation of the animal. After the breeding season is passed, the period of moulting commences. The effects of this exhausting process, which, if not a disease, is closely allied to it, are well known. When the bird recovers its strength, we find it in a new garb, which advances to perfection in proportion to its necessities ; those which migrate to great distances standing in need of a speedy maturation, whilst others continue in the act of moulting between three and four months."

"The most perfect state of plumage is observed in the Spring. Now, if we admit the fact of a vernal moulting, then must this moulting be characterized by other circumstances than those which obtain in the autumnal ; for, after the latter, the plumage requires several months to arrive at maturity ; and the bird, in ridding itself of its excretions, finds itself in too exhausted a state to perform the functions of propagation. The spring moulting therefore, so far from exercising any debilitating influence upon the physical powers of the bird, should seem to afford them additional energy ; for this moulting is pretended to take place about the period of the sexual union, when all the powers of nature are in full vigor."

" In those singing birds which winter with us, we can perceive no diminution of vital energy during the vernal season, either as respects vigor of body or capacity of voice." Referring to the species under consideration the same author continues : "The *Fringilla* (*Chrysomitris*) *tristris*, though migratory, frequently continues the whole year with us ; and his song, in the month of March, while yet his autumnal dress continues, is tuneful and animated. The change in his garb begins to appear in April, and early in May we behold him in his brilliant yellow plumage, which may be termed his bridal garniture, for shortly thereafter commences the period of nidification. During all this season of animation, his tuneful powers are unabated. In September, both sexes are nearly alike, for then they have moulted." (*Familiar Science*, October, 1879.)

The month of August has been pronounced by some bird observers, as the voiceless month of the year. How this fancy has obtained I know not. It is true that the full gush of melody from our feathered guests is over, but it is by no means hushed. While the note of the Goldfinch differs at this season from the swelling note of the vernal period, I have still heard them pour forth in numbers as sweet as those of May. The song of the male bird is clear and precise, if not loud, and is rendered with a varying modulation, especially in the pairing season, similar to that of the domesticated Canary-bird. The female gives utterance to a soft acquiescing note in answer to the strain of her well-beloved, the salute and response of the happy pair, forming in connection, a pleasing medley of bird language.

The flight of this bird is somewhat irregular, consisting of a series of risings and fallings, the latter rather carelessly made with closed wings, the onward course being somewhat undecided and capricious as they seek from right to left and from left to right the seeds they love so well. During the flight, too, they utter over and over again a series of short notes, beginning with strong emphasis and closing with modulated voice. When they have assembled after their family cares for the season have ended, and thus scour the roadsides and fields for a regalement upon the seeds of the season, their note has seemed to me of a somewhat mournful cast, as if full of regret for the days which have passed, and brimming with good-byes to the scenes and associations which have clustered around their homes, and witnessed the fledging of their children.

Mention has been made of their familiarity. This trait of the Goldfinch is of so prominent a nature, that it should not pass without notice. Advantage, too, is taken of this characteristic, to the bird's

own misfortune. So utterly heedless, or careless do they seem to the
approach of man, especially when feeding, that they may be easily
snared with a horse-hair loop attached to the end of a fishing rod. I
have in this way, out of sheer love for the sport, taken many, releasing
them of course after having effected my captures. These captures, too,
have been effected upon the same individual for a second and even a
third time, without much apparent trepidation or care on the part of
the ensnared. So prone is this bird to leave the nutty seeds of nature's
bounty in some favored spot, that when taken therefrom in the manner
already noted, and released, instead of flying away, it will immediately
return thereto. These facts I have noticed and experienced under very
favorable conditions and with a tempting spread at hand. I do not
assert, however, that this bird never evinces timidity. It is when feeding
in companies and upon the weeds that it may most easily be approached
and captured without much caution on the part of the captor. Attempts
have been made to cross this bird with the Canary pet, but I think with
no very pronounced success, although Audubon says that it has, in
certain instances, been done with very satisfactory results. It bears
confinement well, especially so if in company of its own kind, and would,
no doubt, make a very agreeable and pleasing cage-bird, were it not so
common with us as to render such a measure if not very desirable, certainly without profit to bird-fanciers.

Nature occasionally exhibits one of her abnormal phases, in the
albinism of this finch. While this disease, if it may so be called, is of
rare occurrence in any of the birds, it seems to assert itself more
frequently in some families than in others. The family of the *Frin-
gillidæ*, of which the species under consideration is a member, according
to Mr. Ruthven Deane, who has devoted considerable attention to this

interesting subject, seems to furnish albinistic representatives more largely than any other family. A pure white American Goldfinch is in the possession of Mr. Gilbert, of Penn Yan, N. Y. This Finch is, though apparently a delicate bird, strong and hardy enough to brave the winters of this section. While the larger portions of the bands which have assembled early in autumn proceed in October to more southerly localities, many may occasionally be found with us in winter, gleaning a subsistence from the seeds of the *compositæ*. In Dutchess county, N. Y., I have found them very common in winter, associating with the Red-polls, Snow-flakes and Blue Snow-birds. A writer in *Forest and Stream and Rod and Gun*, over the initials M. C. H., and under date of Warner, N. H., May 26, 1878, says : "I have noticed for many years that they (the Goldfinches,) winter here and are often seen in large flocks. During the winter of '76-'77 there was an unusually large flock of these birds in this vicinity, and I often saw them and have specimens taken at the time." Under date of February 2d, 1880, the same writer, regretting a dearth of the feathered tribe for that winter continues : "There have been very few Yellow-birds about, and almost none of late." A great many of these birds passed the winter of 1880 in the vicinity of Fishkill-on-the-Hudson, (Dutchess County, N. Y.,) according to the valuable list of the birds of that locality, by Mr. Winfrid A. Stearns. Mr. H. D. Minot, of Boston, Mass., says they pass the winters about that city.

In this section, the flocks of the Yellow-bird begin to dissemble about the middle of May. Even at this interesting period of their existence, their social habit occasionally asserts itself with energy, and especially may small groups of males be seen chattering good-naturedly, and pluming their satin habits with apparent pride and satisfaction,

while their busy wives are occupied with their home affairs. It would
not be very far from the truth, to pronounce the American Goldfinch a
persistently gregarious bird at all times and places About the middle
of June, though seldom at an earlier period, more often later, the happy
couples are ready to receive: their little neatly constructed dwellings
are finished, their household affairs are wholly regulated, and their
family responsibilities commence. The courtship and domestic arrangements of this bird are so admirably set forth by Mr. Ernest Ingersoll,
that I cannot refrain from quoting at length from that high authority;
he says:

" We ought to be well informed in regard to its breeding habits
(as indeed we are); but in looking into the subject, I find several gaps
in our knowledge—or, more accurately, perhaps, I find gaps in *my*
knowledge, which I am in hopes of having filled by observers who may
care to read a brief review of present information upon the subject."

"Of the courtship of the Goldfinch Mr. Thos. Gentry paints a very
pleasing picture.

* * * * * * * * *

I quote 'In the month of April the flocks dissolve into small parties
preparatory to mating. It is quite common to see two males and one
female together, the former lavishing the most endearing attentions
upon the latter, and besides occasionally regaling her with the most
delicious melodies. Whimsical and exceedingly variable, she selects
one suitor and almost the next moment discards him for the other,
which at this moment is perched near by, pouring out his love in the
most charming manner. This condition of things lasts during a couple
of days before the final choice is made. It seems to require the utmost
condescension, as well as the greatest effort, for the successful suitor

to retain his hold upon her affections, for she is likely to waver in the
interval of time which elapses before nidification is begun. This duty
so completely engrosses her time and attention that the tendency to
flirtation, so to speak, has not time to manifest itself, and is soon
abandoned.'

'Subsequent to mating, and just before nidification, the successful
mate and his partner ramble together in quest of food. When weary
of this business they may be seen perched upon a common twig, when
the former with his sole energy pours out his passion in the most charming language, ever and anon turning toward the object of his love as if
to ascertain whether it meets her approval or not. A soft, low note,
which may be expressed by *taw-yah*, is her sign of recognition.'"

*   *   *   *   *   *   *

Mr. Ingersoll continues : "The site selected for the nest is a tree or
stout bush, in an upright crotch of which, or among supporting twigs
that sprout from a horizontal branch, perhaps far from the trunk, the
nest may be securely fastened. Apple, pear, maple and birch trees,
willow, and other thornless shrubs, are the most common choice. Mr.
Gentry asserts that in his region it requires six days to complete the
structure, and that 'oviposition commences on the ensuing day.' Dr.
Brewer saw a nest built and one egg laid in half that time in Boston,
while a Michigan correspondent alleges two weeks as necessary there.
Probably circumstances alter cases. The female appears to be the sole
artificer. Sometimes dire destruction, in the shape of a gale of wind,
or otherwise, overtakes the half finished homes of these birds and
wrecks all their labor ; but they will courageously rebuild, as witnessed
by Mr. J. P. Hutchins in Central New York ; and Gentry tells of a pair
renewing the attempt to erect their nest in the same place after being

baffled four times." — (*Forest and Stream and Rod and Gun. Vol. X., p.
443. July 11, 1878.*)

THE NEST AND EGGS.

Of the nest and eggs of the American Goldfinch, the same pleasing
writer observes : " Few specimens of ornithic skill in architecture have
been more elaborately dwelt upon than the nest of the Goldfinch. Yet,
although very attractive in its result, it is by no means a conspicuous
example of a bird's ingenuity, as in the well woven purse of the
Baltimore Oriole ; for the Goldfinch, simply mats her pretty materials
together by movements of her feet and body, not attempting to inter-
weave much or knit by the aid of the bill."

" Not being particular as to kind, so that the requisite softness and
pliability are obtained, and gathering materials close by, the various
substances entering into the composition of half the Goldfinch's nests
collected would make a long catalogue. Outwardly, the ordinary nest,
which is about the size of a large tea-cup (only in most cases much
higher), exhibits a felting of vegetable fibres, shreds of reddish bark,
fragments of ragged grasses, leaves, hemp, bits of fungus, tassels and
flowers of various delicate weeds and grasses, with more or less
vegetable wool, spiders' webs and lichens loosely attached. Through
the surface, in such a way as to hold it stoutly in place, pass the
supporting twigs of the crotch in which the whole rests. The mass of
these materials, in which often a great deal of wool, fern-down and the
like stuff is mixed, causes the walls to be thick and dense. Interiorly
a receptacle for the sitting bird is hollowed and lined with fine rootlets,
horse-hair, 'plumose appendages, or pappus of the seeds of composite
plants,' raw cotton and fern-wool. In all these nests one element is

sure to be present—thistle-down. The Goldfinch adorns the walls of her boudoir with its glistening silk, and makes her bed of the elastic gossamer that floats through the summer air."

"As I have remarked, the date of egg-laying varies greatly. It seems to occur earliest on the Southern Pacific coast, late in May; at Sacramento and in Utah about the middle of June; at Trenton, N. J., Dr. Abbott tells me he found their eggs from May 17 to August 3; Gentry puts the time at Philadelphia as 'generally from the 10th to 15th of June;' in New England, Samuels gives June 10, as the earliest date, while Dr. Brewer says it is usually past July 10, before their nests are constructed, and September before the broods are able to fly; in Michigan, eggs are recorded from May 20 to September 25, the first week in August being regarded at the height of the season. It is evident, then, that though the Goldfinch breeds late, as a rule, yet sometimes it nestles quite as early as the majority of birds. Upon this point we need more observations. That the same individuals may and do vary greatly in the time from year to year I have no doubt; why, it is impossible, perhaps, to guess. That they have the power of retaining their eggs, or rather of repressing their desire to lay to a much greater extent than is supposed to be the case with most birds, is proved not only by the long delays which have been known to take place in their nest-building, with a successful *finale*, but also from the fact that those specimens dissected in April show an equal readiness and development of ovaries and testes with those shot late in the summer. I doubt whether anything more rational than caprice can be assigned as the cause of their anomalous irregular breeding: the usual explanation, scarcity of proper food for the young in early summer, has no supporting evidence in what we know of the bird's diet, and is distinctly proved

of no weight by the fact that frequently some Goldfinches do nest early,
while others postpone it until very late, and this in the very same
district."

" The eggs of this species are five, and often six, in number, and in
color faint bluish-white immaculate, the blue tint appearing less strongly
in empty specimens. Measurements of an average clutch were : 66 x .51,
64 x .50, 63 x .50 and 62 x .50 of an inch. Occasionally, as noted by Dr.
Abbott and others, spotted examples are seen—the markings faint and
scarce—but this is very rare and exceptional. The elaborately blotched
eggs described by Wilson, Richardson and some other early authors,
are surely erroneously identified."

" About ten days appears to be the period of incubation. As no
second brood (invariably ?) is to be anticipated, the young are diligently
attended by both parents, who exhibit the most clamorous distress whenever danger threatens. Cases are recorded of their returning not only
to the same locality two or three successive summers, but even constructing a new nest upon the foundation of the old. I have seen such
a structure, and observed not only that the height was twice its diameter,
but that the materials of the second were precisely similar to that used
in the first nest. It is said, also, that this species will bury under a
' second story ' the parasitic egg of the Cow-bird, when it is so unlucky
as to have one thrust upon it, as does the other Yellow-bird, *Dendraca
Æstiva.* When almost a fortnight of age, the young leave the nest,
and soon the little family groups combine into the merry flocks that we
see gaily playing about the sere reeds in autumn, or drifting away in a
cloud of thistle-down before the October breeze." — *Ibid.*

Little can be added, from my own observations, to the foregoing
account. I have, however, been frequently interested and amused at

seeing the female bird, when engaged in nest-building, perched upon a swaying clothes line, near a knot or raveled end thereof, bobbing backward and forward to preserve her equipoise after the manner of the sand-piper, and industriously gathering fine shreds of the line with which to furnish her nest. I have seen the same individual return time and again to the same spot, on each occasion accompanied by her liege and his cheering medley, for the soft fibres she seemed to love so well. When hemp and cotton lines have been strung side by side, each having frayed ends and knots, I have invariably noticed that the former is preferred, owing no doubt to the superior softness of the cotton staple when compared with the hemp.

Sam has placed at my disposal a nest and clutch of eggs of this bird, from his collection. The nest was found in the forks of a willow bush, six or seven feet from the ground. It is secured at its edge to branching twigs, five in number, by a few shreds of bark, an occasional straw, and filaments of weeds. On some of the twigs the latter material extends around them to the lowest base of the structure. On the largest one which is more or less exposed, I notice that the fine cotton ravelings extending from the body of the nest, seem to be fastened thereto by a mucilaginous secretion, without surrounding the twig. A few bits of paper are found near the exterior surface of the nest. Its bottom is $1\frac{1}{2}$ inches thick ; the side walls vary from $\frac{3}{8}$ to $\frac{1}{2}$ inches in thickness. Greatest outside diameter, 3 inches ; inside diameter $1\frac{1}{2}$ inches ; concavity $1+$ inch in depth and lined with seedless thistle down, whole vertical height, 3 inches.

The eggs, four in number, are of the normal hue, without specks, and measure as follows : No. 1, 0.62 by 0.49. No. 2, 0.66 by 0.50. No. 3, 0.68 by 0.50. No. 4, 0.65 by 0.50.

BRIGHT FEATHERS

OR

Some North American

BIRDS of BEAUTY.

By Frank R. Rathbun.

*Illustrated with Drawings made from Nature, and carefully
Colored by Hand.*

AUBURN, N. Y.
PUBLISHED BY THE AUTHOR.
1882.

PLUMAGE, ETC.

The plumose colorations of this species are so generally well known, as to hardly warrant the space for their detail. Many people, however, are exceedingly prone to confound the Goldfinch with the Summer Warbler or Summer Yellow-bird, (*Dendroeca æstiva.*) To such, and to whom a Yellow-bird is a Yellow-bird, be he Goldfinch or Warbler, the detail may be of interest and serve to dispel the illusion. I may say in this connection, that I trust the critical ornithologist will take no exception to this endeavor to enlighten, and pardon the comparison of facts to him so well known.

The difference between the two species is easily recognized if a little critical attention be bestowed upon them. The Summer Yellow-bird may always be known from the brownish spots upon its yellow breast. In both sexes, too, of the species, there is an *entire* absence of any black color. In the male of the Goldfinch much of black obtains, while in both male and female there is an entire absence of markings on the breast ; (referring to mature and normal individuals.)

The predominant color of the male American Goldfinch, is, in summer, of a bright, clear lemon yellow, which blends or fades in many instances, into a pure white on the tail-coverts. The head is ornamented with a well defined jet black cap. The wings and tail are also black, the former edged and crossed with white, and the feathers of the latter spotted at their interior ends also with white. The lesser wing coverts are yellow, and the bill and feet are of flesh color. From the change in plumage in September until the following April, the yellow coat of the male is substituted for one of flaxen brown, paler or whitish below, and the black head-dress is cast aside.

The dress of the female is similar to that worn by the male in winter, having, however, rather more of an olivaceous shade above, and dingy yellowish below. She has no black cap, and the wings and tail are dusky instead of black, with whitish markings.

The bill is rather robust and cleanly pointed ; tail, deeply emarginate and about 0.8 of an inch longer than the tips of the closed wings. The length of this bird varies from 4.50 to 4.75. Alar extent, 8.00 to 8.03. Wing about 2.75. Tail 2.00 inches. Inhabits North America generally.

AUDUBON ON THE AMERICAN GOLDFINCH.

CARDUELIS TRISTIS, LINN.

Plate clxxxi.—MALE AND FEMALE.

" This species passes over the State of Louisiana in the beginning of January, and at that season is seen there for only a few days, alighting on the highest tops of trees near water courses, in small groups of eight or ten, males and females together. They feed at that period on the opening buds of *maples*, and others that are equally tender and juicy. In the month of November they are again seen moving southwards, and for a few days only."

" A few breed in Kentucky and the State of Ohio, but the Middle Districts, are their principal places of resort during summer, although they extend their migrations to a high latitude. They arrive in the State of New York, about the middle of April ; and as they become very abundant in that state during the summer, I shall describe their habits as observed there."

" The flight of the American Goldfinch is exactly similar to that of the European bird of the same name, being performed in deep curved lines, alternately rising and falling, after each propelling motion of the wings. It scarcely ever describes one of these curves without uttering two or three notes whilst ascending, such as its European relative uses on similar occasions. In this manner, its flight is prolonged to considerable distances, and it frequently moves in a circular direction before alighting. Their migration is performed during the day. They seldom alight on the ground, unless to procure water, in which they wash with great liveliness and pleasure, after which they pick up some particles of gravel or sand. So fond of each other's company are they, that a party of them passing on the wing will alter its course at the calling of a single one perched on a tree. This call is uttered with much emphasis. The bird prolongs its usual note, without much alteration, and as the party approaches, erects its body, and moves it to the right and left, as if turning on a pivot, apparently pleased at showing the beauty of its plumage and the elegance of its manners. No sooner has the flock, previously on the wing, alighted, than the whole party plume themselves, and then perform a sweet little concert. So much does the song of our Goldfinch resemble that of the European species, that whilst in France and England, I have frequently thought, and with pleasure thought, that they were the notes of our own bird which I heard. In America, again, the song of the Goldfinch recalled to my remembrance its transatlantic kinsman, and brought with it, too, a grateful feeling for the many acts of hospitality and kindness which I have experienced in the ' old country.' "

" The nest also is perfectly similar to that of the European bird, being externally composed of various lichens fastened together by

saliva, and lined with the softest substances. It is small and extremely
handsome and is generally fixed on a branch of the Lombardy poplar,
being sometimes secured to one side of a twig only. I have also found
it in elder bushes, a few feet above the ground, as well as in other trees.
The female deposits from four to six eggs, which are white, tinged with
bluish, and marked at the large end with reddish-brown spots. They
raise only one brood in a season. The young follow the parents for a
long time, are fed from the mouth, as canaries are, and are gradually.
taught to manage this themselves. When it happens that the female is
disturbed while on her nest, she glides off to a neighboring tree, and
calls for her mate, pivoting herself on her feet, as above described.
The male approaches, passes and repasses on the wing at a respectful
distance from the intruder, in deeper curves than usual, uttering its
ordinary note, and when the unwelcome visitant has departed, flies with
joy to his nest, accompanied by the female, who presently resumes her
occupation."

"The food of the American Goldfinch consist chiefly of seeds of the
hemp, the sun-flower, the lettuce, and various species of the thistle.
Now and then, during winter, it eats the fruit of the elder."

"In ascending along the shores of the Mohawk river, in the month
of August, I have met more of these pretty birds in the course of a
day's walk than any where else ; and whenever a thistle was to be seen
along either bank of the New York canal, it was ornamented with
one or more Goldfinches. They tear up the down and withered petals
of the ripening flowers with ease, leaning downwards upon them, eat off
the seed, and allow the down to float in the air. The remarkable
plumage of the male, as well as its song, are at this season very agreeable ; and so familiar are these birds, that they suffer you to approach

within a few yards, before they leave the plant on which they are seated. For a considerable space along the Genesee river, the shores of Lake Erie, Lake Ontario, and even Lake Superior, I have always seen many of them in the latter part of summer. They have then a decided preference for the vicinity of water."

"It is an extremely hardy bird, and often remains the whole winter in the Middle Districts, although never in great numbers. When deprived of liberty, it will live to a great age in a room or cage. I have known two instances in which a bird of this species had been confined for upwards of ten years. They were procured in the market of New York when in mature plumage, and had been caught in trap cages. One of them having undergone the severe training, more frequently inflicted in Europe than America, and known in France by the name of *galerien*, would draw water for its drink from a glass, it having a little chain attached to a narrow belt of soft leather fastened around its body, and another equally light chain fastened to a little bucket, kept by its weight in the water, until the little fellow raised it up with its bill, placed a foot upon it, and pulled again at the chain until it reached the desired fluid and drank, when, on letting go, the bucket immediately fell into the glass below. In the same manner it was obliged to draw towards its bill a little chariot filled with seeds ; and in this distressing occupation was doomed to toil through a life of solitary grief, separated from his companions wantoning on the wild flowers, and procuring their food in the manner in which nature had taught them. After being caught in trap-cages, they feed as if quite contented ; but if it has been in spring that they have lost their liberty, and they have thus been deprived of the pleasures anticipated from the previous connexion of a mate, they linger for a few days and die. It is more difficult to procure a mule brood between our species and the Canary, than between the latter and the European Goldfinch, although I have known many instances in which the attempt was made with complete success."

"The young males do not appear in full plumage until the following spring. The old ones lose their beauty in winter, and assume the duller tints of the female. In fact, at that season, young and old of both sexes resemble each other."

"There is a trait of sagacity in this bird which is quite remarkable, and worthy of the notice of such naturalists as are fond of contrasting instinct with reason. When a Goldfinch alights on a twig, imbued with bird-lime expressly for the purpose of securing it, it no sooner discovers the nature of the treacherous substance, than it throws itself backwards, with closed wings, and hangs in this position until the bird-lime has run out in the form of a slender thread considerably below the twig, when feeling a certain degree of security, it beats its wings and flies off, with a resolution, doubtless, never to alight in such a place again; as I have observed Goldfinches that had escaped from me in this manner, when about to alight on any twig, whether smeared with bird-lime or not, flutter over it, as if to assure themselves of its being safe for them to perch upon it."

"This interesting species is found on the shores of the Columbia river. It is mentioned by Dr. Richardson, as visiting the Fur Countries, where it arrives at a very late period, as it retires in September, after a stay of less than three months. The eggs described by that most zealous naturalist agree in every particular with some now before me, which I collected myself. They measure a trifle more than five and a half eighths in length, by four and a half eighths in breadth, and are very obtuse at one end and sharp at the other."

THE OOLOGIST.

DENDRŒCA ÆSTIVA, (Gm.) Baird.

THE SUMMER WARBLER. ♂ MALE. ♀ FEMALE. Life Size.

DRAWN FROM NATURE BY FRANK R. RATHBUN

BRIGHT FEATHERS.
PLATE IV.

THE SUMMER WARBLER.

DENDRŒCA ÆSTIVA, (GM.) BAIRD.

PLATE IV.—MALE AND FEMALE.

PRING opens wide the gates of her teeming aviary for the egress of a no more familiar or confiding bird than the subject of the present sketch. Named for the warm-blooded sister of the seasons who succeeds her, the Summer Warbler comes to us upon the laggard train of her vivifying skirts, seldom later than the first days of the month of May. On special occasions, however, when her frigid brother has retired and yielded up his crystal scepter before her redolent breath earlier than his usual wont, this cheerful bird may be heard giving vent to his sweet refrain in the elder days of April.

ORIGINAL INITIAL.

In this, the vernal period of the year, when the young leaves are striving to burst the barriers of their scaly capsules, and laboring hard for freedom and the light that they may anon subdue with grateful shade the glinting shafts of a searching sun, the restless and symmetrical

form of our welcome guest may be seen surveying his natural surroundings with a curious and searching eye.

He comes to us as the breeze bids him. His advent has been
heralded by his well trained tongue in the meridian of a spring time
day, often before his graceful form has been noted. In the gray dawn
of an awakening sun, his cheerful tune has caught the ear for the first
time in the budding period, and, alas! he has startled us with his crisp
reveille while we were hugging a wooing pillow and knew not whether
of the twain we had dreamed or actually heard his good morrow.

We involuntarily touch our visor to the Summer Warbler, the reddish spots upon his golden livery suggestive of the chevrons they faintly
outline, as the trim sergeant of a squad of days that come trooping on,
full of promise of the sun, of blossoms, and of the droning chorus of
insects. With his advent, the days are marshaled which steep the
senses in a languid, feverish unrest, and incite the gypsy strain which
tincts the blood of our primitive natures to its assertion. The days
have come, in which we eye our gun askance, when we assort and
arrange our flys, our snells, leaders, hooks and lines, and joint our rod
to test its supple strength as we yearn for the greenwood and the brawl-
ing brooks. Days, when the earth emits a life-giving odor, and its
garniture of innocent violets yields up their fragrance as incense to the
pregnant goddess of the seasons. Days, when the sky seems of opal,
and the clouds of pearl, the light of gold, and the rain-drops of silver.
Days, when the ardent and long forgotten loves of our boyhood assert
their empire, and we dreamily wonder if the locks and lashes we so
madly adored, slumber in the dust or sweep over joys maternal. Days,
with which the Summer Warbler comes, and in whose voluptuous
zephyrs the very shrubs upon the lawn and the "pussy willows," beside

the burly brooks, seem to shiver with a mysterious joy as they feel the
sap, their life blood, well up into and throughout their varied members.
Days, when we can imagine that the leaves have voluntary motion, and
seem to clap their glad palms together in a round of applause, throughout every aisle and avenue of their sinuous assemblage in recognition
of the jubilant carolings of the beautiful and innocent denizens of their
shady naves. Days, which fret the links upon our longing limbs that
bind us to duty and labor, and entrance the child of nature, and days in
which we murmur against the heritage bequeathed us by the frail one of
Eden. We are recompensed, however, for the wearing of the shackles
which restrain our eager feet and bar our entrance into nature's realm,
in the joyful notes which fall from the tongues of the birds. In the
sweet medley which they pour forth, the strain of the Summer Warbler
is by no means insignificant. Once heard, and its source noted, it will
never be forgotten by the admirer of nature's music. Its syllabic crisp-
ness commands the ear, and its hearty and confiding utterance wins from
the heart of the listener, a love, involuntary, sincere and faithful which
it were vain to withhold.

Simple, confiding and sweet, I imagine the song of this warbler may
be best represented by an expression from between the nearly closed
lips and teeth, of the syllables, *sin-sin-sis-a-rai*, or *sin-sin-sin-sisa-rai*,
with the accent and rising inflection upon the last syllable. The bird
seems most happy in its repetition, hardly ceasing to give it utterance
even in the swelling noontide of a summer day.

Owing to a not too close distinction of plumose colorations and
markings, I find as previously remarked that many persons are apt to
confound the Goldfinch described in the preceding sketch, with the bird
under consideration. I have thought proper, therefore, even at a sacri-

fice of variety in color in "*Bright Feathers*," to let a description and
delineation of the one succeed the other, in order that an intelligent
comparison might be made between the two species.

Like the Goldfinch, the Summer Warbler has been decorated with
a variety of appellations, both local and scientific, among which, those
of the Summer Warbler and Summer Yellow-bird, appear to be most
generally recognized and most common. By various authors he has
been christened the Golden Warbler, Yellow Warbler, Yellow-poll
Warbler and Blue-eyed Yellow Warbler, any of which do him full justice
and give an intelligent conception of his colorations. Of the various
names given him by eminent ornithologists, and which are believed by
Dr. Coues, to be strictly applicable to the North American bird, we
select a few as follows : Olive Warbler, *Pennant* ; Yellow Titmouse,
Catesby ; Citron Open-bill, *? Raf* ; Citron Warbler, *Swainson & Richardson* ; Children's Warbler, Rathbone's Warbler, *Audubon* ; and Yellow
Wren, *or* Willow Wren, *Nuttall.*

This charming Warbler, no less famous for its beauty and song than
for its excellent character and confiding disposition, is one of the most
numerous representatives of the family of warblers that favor us with
their welcome presence and delight us with their sweet notes. Any
time after the first of May, and until the latter part of September, it
may be found flitting amid the green of the trees in the very heart of
our most populous districts. It seems, indeed, to prefer the shrubbery
of our lawns, the orchard and garden in which to build its nest, being
one of the very small number of the genus that are so inclined. Its
continuous activity I have remarked as one of its most prominent
characteristics. It seems continually searching, feeding or nest building, without intermission or rest throughout the live-long day. Its food

consists largely of aphides, grubs and larvae, which it picks out from the interstices of the bark of the trees, the crevices of the fences and from the edges of the lap-boards of the buildings. It catches insects, too, upon the wing, with a grace and dexterity equal to that of the fly-catchers, and frequents the currant bushes and vegetable gardens in search of worms and the eggs of insect pests. Throughout the round of his expeditions he never forgets to sing ; he cheers himself along if no one else. Especially when in the trees does he indulge in his sweetest notes. Throwing his whole bulk to the front, with mandibles thrown open and upward, the notes trip off his nimble tongue with a flourish that almost knocks him over, and he is away soon to reappear and repeat his jolly carol.

A very prominent characteristic of the Summer Warbler, and one which has excited universal comment and admiration from those who are conversant with its habit, is the resolution it displays in refusing to incubate the alien eggs of the Cow-bird, by whom this bird is frequently victimized. Seemingly too indolent to speed to her own home, the female of the Cow Black-bird very frequently avails herself of the nests of other and smaller birds in which to deposit her eggs, thus ridding herself of the cares of incubation and provision for her helpless progeny.

With persistent determination, however, the Summer Warbler refuses the additional burden, and with remarkable sagacity makes shift to "avoid the hateful imposition even to the length of sacrificing its own eggs and giving up its nest." When disinclined to proceed to this extremity, it will rebuild a new structure immediately over and upon the original one, burying beneath it the alien egg, and such of its own as may be therein, and proceed as it originally intended. Should a second trespass be committed on the new nest, it will frequently repeat

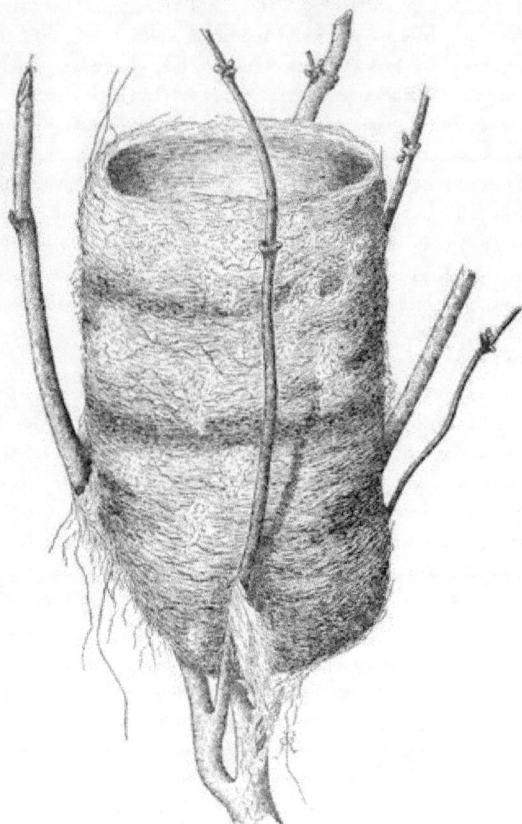

FIG. 1.—THREE-STORY HOME OF THE SUMMER WARBLER. ABOUT ACTUAL SIZE.

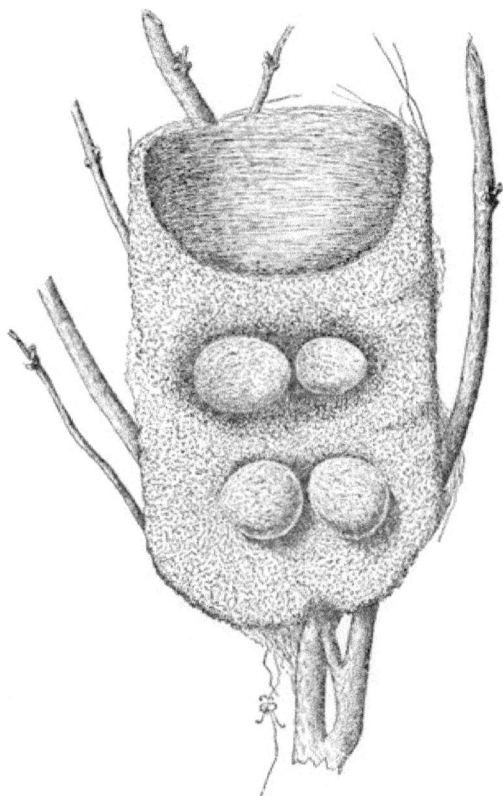

FIG. 2.—VERTICAL SECTION OF THREE-STORY HOME OF THE SUMMER WARBLER.

its labor and bury its contents also, by constructing even a third nest over the second, resulting thus, in a three story structure, which affords a wonderful example of the little creature's sagacity and perseverance, and a curiosity in bird architecture.

For several years I have had in my possession a nest of this description, which was found in a honey-suckle vine but a few feet from the ground. I have sacrificed the structure for the purpose of making the sectional illustration shown in Figure 2, p. 71. Figure 1, p. 70, is made as nearly the size of the original as possible, although owing to the length of time in which I have retained it, it may have contracted or settled somewhat in its height. This fabric was composed of the usual material, as will be described further on, and upon making a careful vertical bisection with a pair of scissors, revealed an internal arrangement as shown in Figure 2. In the bottom compartment or cellar of this unique home, was found two eggs of the Cow Black-bird relatively opposed to each other as I have drawn them. A slight accumulation of short, fine, fibrous material appeared to have been cemented between and over the obnoxious eggs before the new or secondary structure was begun. In the second story was found two eggs, one of the Cow-bird, and the other the legitimate egg of the " Mistress of the Manse." Both of the latter eggs were completely surrounded by an aggregation of short material, so firmly cemented together, that it was somewhat difficult to cut through the mass. The dark shading which surrounds the eggs in the second compartment or nest faithfully represents the situation. Over the second nest, a third had been constructed which was empty when the specimen was taken. As I have no notes or dates of any kind relating to the find, save that it was taken from a honey-suckle vine as already mentioned, it is only

BRIGHT FEATHERS

OR

Some North American

BIRDS of BEAUTY.

BY FRANK R. RATHBUN.

*Illustrated with Drawings made from Nature, and carefully
Colored by Hand.*

AUBURN, N. Y.
PUBLISHED BY THE AUTHOR.
1882.

reasonable to surmise, at least to hope, that the persecuted bird had through all these vicissitudes of vexation and labor at last been successful in rearing its own legitimate brood without further imposition.

TWO-STORY BIRDS' NESTS.

By the courtesy of the *Scientific American*, I am enabled to present to my readers the foregoing beautiful illustration of a two-story home of the Summer Warbler, engraved from an original drawing made by Mr. Daniel C. Beard.

Mr. Beard's very happy account of this Warbler and its famous characteristic, is of too interesting a nature to be omitted. He says :

" While the expanding leaves of tree and shrub retain the tender tints of pink, and the broad lily pads commence to mosaic the surface of the ponds with green, in perfect harmony with the bursting bud and opening flower comes the summer yellowbird, and from hedge and bush may be heard his song, as simple and pleasing as the tasteful but modest plumage that covers his little person. Almost immediately after the first appearance of these industrious little birds they commence their preparations for housekeeping. The male bird flies busily about selecting such material as feathers, plants, fibers, the furze from ferns, the catkins from willows, and other similar objects, all of which he brings to his mate, who arranges and fashions their delicate nest. So quickly and deftly do this little couple labor that they build the greater part of their house in a single day."

" There is often a third party interested in the construction of this nest, a homeless, happy-go-lucky Bohemian bird, who has a sort of tramp's interest in the housekeeping arrangements of most of the smaller feathered denizens of copse and woods. This is the well-known cow blackbird, who disdains to shackle her freedom with the care of a family and shifts a mother's responsibility by farming her progeny out, while she seeks the incongruous but apparently congenial companionship of the cattle, with whom she appears to be on the most intimate terms."

" The cow blackbird deposits its eggs indiscriminately among the nests of smaller birds. The blackbird's eggs generally hatch out a day or two before the adopted mother's own eggs, so, when the legitimate members of the family do come, it is to find their nest already occupied by the strong, lusty interlopers, who, on account of their superior size

and strength, come in for the lion's share of all the food brought to the
nest. Thus the innocent parents rear the aliens, while their own young
starve. It is really a pitiable sight to see a couple of little greenlets
anxiously searching from daybreak till evening for food to fill the
capacious crop of one or more young cow blackbirds, considerably larger
than the greenlets themselves."

" The summer yellowbird, though confiding little creatures, are not
readily duped or imposed upon. Their instinct is sufficiently near
reason for them to detect the difference between their own little fragile,
prettily-marked, greenish-colored eggs and the great dark colored ones
the vagabond cow blackbird has surreptitiously smuggled into the cozy
nest. The domestic little couple cling to the spot selected for their
house and will not leave it, neither will they hatch the obnoxious eggs,
which they are apparently unable to throw out; but the difficulty is
soon surmounted, and so are the gratuitous eggs, for the indefatigable
workers proceed at once to cover up the cow blackbird's eggs, constructing a new nest on top of the old one, building a second story, as it
were to their house."

" Last summer Mr. Lang Gibson, brought me one of these two-
story nests which he found at Flushing, L. I. ; the lower nest contained
two cow blackbird's eggs, and the upper one three eggs of the summer
yellowbird. Gibson watched the construction of the nest. Visiting it
again after it was finished, he discovered the egg of a cow blackbird.
Next day two of these eggs occupy the nest. Some time afterward, to
his surprise, he found the nest contained three eggs of the yellow-bird
and no signs of the existence of those deposited by the blackbird, but
the nest had the appearance of being much taller than at first, and an
examination disclosed the true facts of the case."

"The accompanying illustration was drawn by the writer from this compound nest. The upper story or nest is partly lifted so as to show the cow blackbird's eggs in the nest below."

" Fig. 1 shows the cow blackbird's egg, and Fig. 2 the yellowbird's egg. These are drawn exactly the size of nature."

" Mr. Nuttall was the first naturalist, I believe, to record the observation of these two story nests. Baird mentions a three-story nest, each of the lower nests containing the eggs of the cow blackbird, the whole structure being seven inches high."—*Scientific American*, -Vol., XLVI.—No. 11. (New Series,) March 18, 1882. p. 169.

Another peculiarity of the Summer Warbler is of such an enigmatical nature, that I am constrained to give in full the evidence of Mr. J. A. - Allen, on this point. Not only this bird, but others of entirely different genera have exhibited the same mysterious behavior. The similar conduct of the common Robin has been noted, and I have witnessed like behavior, in a less degree, in the Cedar, or Cherry Bird.

Mr. Allen, says : " For several weeks the bird (*Dendraca æstiva*) has been in the habit of frequently visiting a grape-vine trellis in front of a window of the dining-room, from which he has been accustomed to sing, wholly undisturbed by the people or the proceedings within the room. Although the trellis has been a favorite resort for the bird, his behavior was not especially noteworthy till June 7, (1879,) when he began to persistently fly against the window-panes, often striking them with considerable violence."

" The trellis stands about eighteen inches from the window, and the portion immediately in front of it is nearly bare, and consists of two horizontal bars, about three feet apart. These form his perch, from which he usually makes his dive at the window. Immediately in

Memoirs of the Nuttall Ornithological Club.

No. II.

THE IPSWICH SPARROW

(*AMMODRAMUS PRINCEPS* MAYNARD)

AND

ITS SUMMER HOME.

By JONATHAN DWIGHT, JR., M. D.

WITH A COLORED PLATE.

CAMBRIDGE, MASS.
PUBLISHED BY THE CLUB.
AUGUST, 1895.

front of the window is an open field with a group of five large apple trees, all within twenty to fifty feet of the house. These, with the trellis and portions of the grape-vine it supports are vividly mirrored in the window, as well as the general landscape, and of course the bird himself whenever he visits the trellis. But his own reflection does not seem to be the point of attraction, as he *usually* strikes the pane two or three feet above the point opposite his perch, but sometimes dives down from the upper bar of the trellis to the lower panes of the window. Occasionally he flies directly from the apple trees against the window, but generally first alights on the bars of the trellis. For several days his visits have begun with early day-break, and have been continued throughout the day till after sunset, he rarely leaving the window for more than a few minutes at a time. He sings almost constantly. I have seen him strike the window-panes as many as ten times in a minute, barely pausing on the trellis between each plunge, long enough to utter with much energy his shrill little song. These proceedings, he will sometimes repeat for several minutes, then fly to the trees and return again a minute or two later, usually with a canker-worm in his beak obtained from the apple trees. This he usually bruises on the trellis-bar and swallows at once before diving at the window, but not unfrequently makes several plunges at the window with the worm in his beak. He strikes the window pane with such force that the clicking of his bill and feet against the glass may be heard to a considerable distance. He usually strikes the large pane a foot or two from the top, fluttering upward to the top, when he returns to his perch. The upper panes receive the chief part of his attention, but he not unfrequently descends to the lower ones, which he follows upward in the same manner to the top of the lower sash. He takes little notice of people standing quietly before the window, and will often strike the pane within six inches of the observer's face."

" If the upper sash be lowered a few inches he will often, after flying against the glass, perch on the top of the open window, peer into the room, utter his song, hop to the trellis, and immediately repeat the operation. I once drew the upper sash half-way down, so as to give him free access to the room. At first he would strike the glass as usual, and then perch on the sash. I left the room for an hour, and on returning found him a prisoner between the sashes, he having evidently in the meantime entered the room, and in trying to make his exit had fluttered down between the sashes, where he had obviously been struggling for some minutes. I freed him, and presumed that this experience would serve to cure him of his strange infatuation for the window. This was on the evening of the first day, but he returned early the next morning to the window, flying against it with unabated persistency. This has continued for three days, and the window seems to have lost none of its charm for him."

" In other respects he seems a perfectly sane bird; he has a mate and a nest in one of the neighboring apple trees, and when it is approached he leaves the window and flies about the intruder with manifestations of extreme solicitude. He is also quite vigilant in driving away other small birds that venture too near his home. Whether he mistakes his own reflection in the window for a rival, or what the charm is, is not obvious, as his behavior in all other respects is apparently entirely natural. As already stated, he almost invariably strikes the window-pane at a point either considerably above or below his perch on the trellis, so that evidently he does not aim at his own reflection in the window. *J. A. Allen, Cambridge, Mass.*"

"P. S.—His visits to the window became less frequent on the fourth day, but were continued with considerable frequency for about ten or twelve days, when the bird wholly disappeared, being caught, it is feared, by a neighbor's cat which had been observed lying in wait for it at the window on various occasions.—J. A. A."—*Bulletin N. O. C.* Vol. iv, pp. 180 to 182, inc.

The Summer Warbler is almost universally distributed throughout North America, and breeds in nearly every part of its extended range. It is a very common visitor in Central New York, and cannot fail to win the love and consideration of those who may feel disposed to cultivate his agreeable acquaintance. Like all the members of the family to which he belongs, he is of inestimable benefit to the agriculturist and fruit-grower, and for this reason as well as for his note, should be encouraged and harbored wherever he makes his appearance. His general character may be embodied in a certificate in which the following words should appear in italics.—*Companionable, innocent, useful and sweet of voice.*

THE NEST AND EGGS.

The nest of the Summer Warbler, like that of the American Goldfinch, is very neat in structure and possessed of durability. The latter quality owes as much to the compactness with which it is built, as to the substances of which it is composed. The materials entering into its composition are generally of a very miscellaneous character, and in a single structure may be found often, horse hairs, soft ravellings from clothes lines, wool and hempen fibres, sometimes silk if it has fallen in the way of the artificer, a few feathers, grasses, or plumes of weeds, and other vegetable down. The finer and softer substances mentioned, are

closely felted together by the tiny feet of the builder for the interior of
the nest ; the coarser fibres forming its exterior frame work.   The nest
may sometimes be found in fruit or shade trees at considerable distance
from the ground ; more frequently, however, in the low bushes, hedge-
rows, and shrubbery of our gardens and fields, and the low-lying tangle
of alder and willow along the banks of streams and other moist situa-
tions.   A nest before me taken from a low bush, about five feet from
the ground, measures as follows : Total depth, two inches ; largest outside diameter, three inches ; outside diameter across the top, two and
one-fourth inches ; diameter of cup, one and seven-eighths inches, and
average depth of cup, one and three-fourths inches.   This little home is
composed principally of dried grasses and silvery filaments of weed stalks
exteriorly.   Its interior of vegetable down, woolly substances and a few
horse-hairs.   The edge of the cup, is finished with a neat firm sort of
selvage of vegetable fibres and horse-hairs interwoven with each other.
The upper edge of the nest from one point of view has a convex, from
another point, (turned one-quarter around) a concave contour for its
outline.   It bulges slightly at its middle-height and conforms at the
bottom to its supports.

The eggs of the Summer Warbler are of a dull grayish or greenish
white, sometimes nearly pure white, indiscriminately dotted and marked
with various shades of lilac and warm brown, the markings being chiefly
disposed around the larger end.   They measure from 0.64 to 0.69 in
length, by 0.48 to 0.53 in breadth.   A clutch of four from Sam's
cabinet measures as follows : No. 1, 0.65 by 0.50 in. ; No. 2, 0.67 by
0.52 in. ; No. 3, 0.69 by 0.51 in. ; No. 4, 0.66 by 0.52 in.

PLUMAGE, ETC.

The plumage of the adult male Summer Warbler, is of a rich golden
or cadmium yellow. The back is of a yellowish olive tint, and is
frequently marked with dark streaks while they may as frequently be
wanting. The breast and sides are boldly marked with lanceolate spots
of an orange brown, which run so closely together as to present from a
little distance the appearance of stripes. The wings and tail are of a
dusky hue, having all the feathers edged with yellow. The crown is of
the same color as the under parts and often marked, especially in high
plumage, with orange brown. Iris, bluish; bill, dark horn-blue; feet, pale
brown. Length, 4¾ to 5 inches; extent, 7½ to 7¾ inches; wing, 2½
inches; tail, 2 inches; bill, 0.37, and tarsus, 0.68 in.

The general coloration of the adult female, is somewhat paler than
that of the male, the yellow olive of the upper parts extending on the
crown, while the spots or streaks of the breast and sides are generally
wanting. The young are still more dully colored than the female. The
upper parts are of an ochrey-olivaceous shade including the crown, and
the lower parts are of a dull yellowish hue, which also obtains upon the
edgings of the tail and wings.

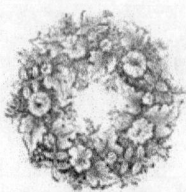

SETOPHAGA RUTICILLA, (Linn.) Sw.

THE AMERICAN REDSTART. ♂·MALE. ♀·FEMALE. Life Size.

DRAWN FROM NATURE BY FRANK R.RATHBUN.   Bright Feathers. Plate V.

THE AMERICAN REDSTART.

SETOPHAGA RUTICILLA. (LINN.) SW.

PLATE V.— MALE AND FEMALE.

R. ELLIOTT COUES, gives such an interesting and graphic account of the notes, food and behavior of this vivacious and beautiful little fellow, and which tallies so closely with what I know of him, that I would be doing the readers of "*Bright Feathers*" an injustice did I not give them the full benefit of his truthful and beautiful description. In writing more particularly of the behavior of the American Redstart, this versatile author and accomplished ornithologist says :

"The Redstart shines among the birds that throng the woods in spring, when his transparent beauty flashes like a lambent tongue of flame at play amidst the tender pale green foliage of the trees. The brilliant little meteor glances here and there in seeming sport, with most exuberant vivacity, as if delighted to display in every action of his

tiny body the full effect of color contrast, shifting every moment into
novel combinations with the cool shade of the background, himself the
foremost figure of an animated picture. But with all this grace and
elegance, this revelry and waywardness, when color plays the pleasing
part of a continual surprise, the Redstart has an eye to business, and
incessantly pursues the gauzy creatures that furnish food to him and all
his kind. You may know him even in his early incompleted dress, and
never fail to recognize his less conspicuous mate, by several characteristic traits. In their unceasing forays on the insect world, they have a
fashion of skipping rapidly along the larger horizontal boughs of trees,
with lowered head and drooping wings, and with incessant sidewise flirting of the fan-shaped tail, that best displays its pretty parti-coloration,
the attitude and action being exactly those you have observed in the
poultry-yard, when the sultan of the harem pursues a reluctant fugitive.
These headstrong raids along the limbs are changed at intervals, when
still more buoyant and more dextrous action absorbs the ceaseless stream
of the Redstarts' energy ; without a moment's pause, the birds shoot out
to this side or to that, and capture insects on the wing in the most
spirited manner ; they dart in zigzag, generally downward, while the
repeated clicking of their mandibles, as turn after turn is executed at
seeming random, yet with admirable precision, tells with what success
these dashing guerillas wage their warfare. Such raids are made right
through the ranks of the airy little insects that swarm in the sunbeams,
and at every descent into their midst not one, but many, of the midges
meet their fate ; the Flycatcher regains his foothold with marvellous
celerity, and races as before along the limb, with many a twitter of
delight, till he is lost to view."

"The Redstart's notes are very curious ; though scarcely describable

they are easily learned, and not likely to be forgotten after they have
been heard a few times; and indeed one may listen to them without the
slightest difficulty, so incessantly are they uttered during the breeding
season. The actions I have endeavored to portray are invariably accompanied by these queer sounds in the intervals between the side-raids
after flying insects. They are rather feeble notes to come from so
sprightly and energetic a performer, though delivered with much
animation and endless repetition. Wilson rendered their ordinary song
by the syllables *weese, weese, weese,* and alludes to several variations of
this twitter his ear had learned to distinguish. '" Many of these tones,"'
says Nuttall, '" As they are mere trills of harmony cannot be recalled
by any words. Their song on their first arrival is however nearly
uniform, and greatly resembles the *'tsh 'tsh tshee, tshe, tshe tshea,* or *'tsh
'tsh 'tsh 'tshitshee* of the Summer Yellow-bird (*Sylvia æstiva*), uttered
in a piercing and rather slender tone; now and then also agreeably
varied with a somewhat plaintive flowing *'tshe 'tshe 'tshe,* or a more
agreeable *'tshit 'tshit e'tshee,* given almost in the tones of the Common
Yellow-bird (*Fringilla tristis*). I have likewise heard individuals warble
out a variety of sweet, and tender, trilling, rather loud and shrill notes,
so superior to the ordinary lay of incubation, that the performer would
scarcely be supposed the same bird. On some occasions the male also,
when angry or alarmed, utters a loud and snapping chirp."' It is
probably to such notes as these last that Wilson alludes in rendering
the sounds by *sic, sic, saic.* Audubon attempts to indicate the sounds in
a still different way. I quote the whole paragraph, which gives a
pleasing glimpse of the bird again. '" It keeps in perpetual motion,"'
he says, '" hunting along the branches sidewise, jumping to either side
in search of insects and larvæ, opening its beautiful tail at every movement

which it makes, then closing it, and flirting it from side to side, just allowing the transparent beauty of the feathers to be seen for a moment. The wings are observed gently drooping during these motions, and its pleasing notes, which resemble the sounds of *teetee-whee, teetee-whee,* are then emitted. Should it observe an insect on the wing, it immediately flies in pursuit of it, either mounts into the air in its wake, or comes towards the ground spirally and in many zigzags. The insect secured, the lovely Redstart reascends, perches, and sings a different note, equally clear, and which may be expressed by the syllables *wizz, wizz, wizz.* While following insects on the wing, it keeps its bill constantly open, snapping as if it procured several of them on the same excursion. It is frequently observed balancing itself in the air, opposite the extremity of a bunch of leaves, and darting into the midst of them after the insects there concealed." '

Dr. Coues, now makes reference to Prof. Gentry's pleasing description of the song of the Redstart, which together with a happy account of this bird's actions and food, will give the reader a clear conception of its maneuverings and *menu.* I quote from Prof. Gentry : "The song of the Redstart resembles very closely that of the Black and White Creeper, but differs in being less prolonged, and in its quicker, sharper, intonation. It may be very appropriately represented by the syllables *tsi-tsi-tree,* the last ending rather abruptly. Its ordinary call is a simple *tsich,* which is heard at long and irregular intervals. Singular to say, these sounds are most frequent when the bird is most active, and not while in the enjoyment of the quietude which follows such a life. At such times our hitherto energetic friend maintains the utmost silence. Being an extremely early riser, it is in the cool, calm hours of the morning, ere all Nature is astir, that he regales the listening spirits of the groves with his sweetest music. About four o'clock he awakes from his slumber, arranges his toilet with care, and with a happy heart starts out to breakfast. But few of his neighbors are up, and for a while he has everything his own way. For nearly five hours he is a busy gleaner. Fastidious in appetite, he does not accept whatever he meets with, but prefers his viands to be of the very best that the great market of the world possesses. While beetles are devoured when other articles are not convenient, there is manifestly a strong predilection for the jucier fly and moth, or the honey-bearing aphis."

An actual examination by Mr. Gentry, (at various times?) of the contents of the stomach of the Redstart, disclosed no less than eight species of coleopterous insects; eight species of hymenopterous and six kinds of dipterous insects; several kinds of aphides and small spiders such as infest the bark, leaves and flowers of plants; mature forms of lepidoptera of various kinds, together with the larvæ of various species and several kinds of cutworm. "Yet no hint of indigestion!" says Dr. Coues, and from this array of fare, he infers we must admit that the Redstart is not only a good hunter, but a voracious and indiscriminate feeder, like some other beauties we may know of. He thinks, too, that Dr. Brewer attests another curious parallel between this bird and other reigning belles: '"Even when lamenting the loss of a part of her brood, and flying round with cries of distress, the sight of passing insects is a temptation not to be resisted, and the parent bird will stop her lamentations to catch small flies."'

" Belonging, as it does," the Dr. continues, " to a semi-tropical group of Warblers, the Redstart would be supposed neither to linger with us during the winter, nor to be among the earlier spring arrivals of the country at large. I have no information of the bird as an inhabitant of

any part of the United States in winter ; on the contrary, at that season
it is present in tropical America as far south even as Ecuador. From
such resorts it moves probably in February, as we hear of its reaching
our southern border at the beginning of the following month. It does
not become generally distributed in this country, however, until some
time in April, becoming numerous in the Middle districts after the
middle of this month, reaching New England and our northern border
about the first week in May, and then soon gaining the limits of its
northward migration. Its movements are quite regular, and at the
height of the season the bird is too abundant in all suitable localities to
be overlooked. The return movement is rather early, all the birds, as
a rule, passing through the Middle districts during the month of September. It is not so common a bird, apparently, in the West as the East,
and the nature of the Rocky Mountain region either interferes with the
orderly north and south movement, or else obscures our recognition
of the periods of migration. It is well known to occur westward into
the Middle Province, but has not been observed in the Pacific slopes.
North, its range is probably nearly coincident with the limit of large
trees ; such extreme of distribution does not seem to be gained until the
latter part of May, and its coming must be immediately followed by
pairing and nesting, as the eggs have been found at Fort Resolution by
the middle of June. While I was collecting at Pembina, on the Red
River of the North, latitude 49°, during the whole month of June the
Redstarts were very abundant in the heavy timber of the river-bottom,
in full song, pairing and nesting, and at the height of their sexual irri-
tability. I never saw it in Arizona, nor have the later students of the
ornithology of that Territory found it, though we have advices of its
occasional appearance in New Mexico, and of its presence in consider-

www.ingramcontent.com/pod-product-compliance
Lightning Source LLC
Chambersburg PA
CBHW021941190326
41519CB00009B/1096